DEADLY
SEASON

DEADLY SEASON

Analyzing the 2011 Tornado Outbreaks

KEVIN M. SIMMONS AND DANIEL SUTTER

AMERICAN METEOROLOGICAL SOCIETY

All photographs courtesy of Chris Karstens/Iowa State University. Photos by Chris Karstens; compiled and edited by Beth Dayton.

Published by the American Meteorological Society
45 Beacon Street, Boston, Massachusetts 02108

The mission of the American Meteorological Society is to advance the atmospheric and related sciences, technologies, applications, and services for the benefit of society. Founded in 1919, the AMS has a membership of more than 13,000 and represents the premier scientific and professional society serving the atmospheric and related sciences. Additional information regarding society activities and membership can be found at www.ametsoc.org.

For more AMS Books, see www.ametsoc.org/amsbookstore. Order online, or call (617) 226-3998.

Library of Congress Cataloging-in-Publication Data

Simmons, Kevin M.
  Deadly season : analyzing the 2011 tornado outbreaks / Kevin M. Simmons and Daniel Sutter.
       p. cm.
  Includes bibliographical references.
  ISBN 978-1-878220-25-7 (pbk.)
  1. Tornadoes—United States. I. Sutter, Daniel. II. American Meteorological Society. III. Title.
  QC955.5.U6S56 2012
  551.55'30973090512—dc23

                    2012002607

  ❀ Printed in the United States of America
  by King Printing Company, Inc. This
  book is printed on recycled paper with a
  minimum of 30% post-consumer waste.

CONTENTS

FIGURES

TABLES

AUTHOR THANKS

The authors would like to thank our editor at AMS Books, Sarah Jane Shangraw, for encouraging us to write this book and for her work with us on this and our earlier book. We also thank Greg Carbin of the Storm Prediction Center for providing us with preliminary tornado reports from 2011. We again thank Jeff Kimpel for encouraging two economists to research tornadoes.

Kevin Simmons
I would like to express my thanks and gratitude to my wife, Susan, for always being my cheerleader.

Dan Sutter
I thank my wife, Natalie, for her love and support; my dogs, Norm, Cliff, and Diane, for sacrificing several daily walks while I was working on the manuscript; and my son, Chuck, for obeying his due date so I could finish the book.

DEDICATION

Kevin Simmons
To my children, Drew and Haley.

Dan Sutter
To my beautiful and wonderful wife, Natalie.

FOREWORD

Prior to 1950 there was no public forecasting of tornadoes in the United States. The average number of tornado fatalities has decreased over the decades since then, a testament to the value of that information and its increasing accuracy and timeliness, which enables the public to take shelter. No tornado had killed more than 40 people in the years 1980 through 2010. In recent years there was some sense that, in the absence of a violent tornado hitting some major outdoor venue, the days of multiscore-fatality tornadoes were largely over. That false hope was shattered in 2011 as violent tornadoes took aim on communities with devastating results. They brought 550 fatalities—the most in the United States since 1936. The tornado that hit Joplin, Missouri on May 22 caused 159 direct deaths and additional indirect fatalities, making it the deadliest individual tornado in the United States since 1947. A pair of Alabama tornadoes on April 27 killed 72 and 64 people, ranking them as the next two deadliest tornadoes in the United States since 1957. With 316 fatalities, April 27, 2011 tied with March 21, 1932 as the second-deadliest day of tornadoes, trailing only March 18, 1925—the day the infamous Tri-State tornado killed 695 Americans (other tornadoes brought the day's death toll to 747).

The authors have statistically examined just about every conceivable factor that might have contributed to the high death toll from tornadoes in 2011, following decades of progress in reducing tornado lethality. Their analysis includes tornado characteristics, demographic and socioeconomic factors, and warning parameters. Many of the deadliest tornadoes during April occurred in the Southeast, and one chapter of the book examines whether that region is more vulnerable to deadly tornadoes. While the latter is found to be true, a more detailed analysis reveals that the killer tornadoes in 2011 often did not follow past tendencies.

This book examines statistics like death rate per million of population to assess whether we are any better off now than in the pre-1950 "stone age" years with no public tornado forecasting. An enlightening part of that analysis includes computing the magnitude of the 2011 deaths as an anomaly relative to recent years, and how it stacks up against the anomaly values for deadly years many decades ago.

The year 2011 has clearly shown that we are still extremely vulnerable to tornadoes individually and as a society. Development of and advances in numerical modeling and prediction of the atmosphere, sophisticated satellites, Doppler radar, and storm-savvy meteorologists have dramatically improved the forecasting and warning of tornadoes and severe thunderstorms. In reality, the advance information available is usually excellent. Of the 550 tornado deaths in 2011, 547 occurred within an area that was under a tornado watch, and two more were within a severe thunderstorm watch. Tornado warnings are issued about 13 minutes in advance of tornado formation, on average, and were often much better than that for the killer tornadoes of 2011. But urban sprawl has made densely populated areas bigger targets than what were once "needle in a haystack" cities. Buildings are not constructed to withstand the winds of strong and violent tornadoes, so without an underground or specially constructed above-ground tornado shelter, there may be no safe place to go even when those warnings are received. Despite the multitude of delivery platforms now available, some people still do not receive the watch and warning information.

The 2011 season may have been a wake-up call regarding the realities of tornado preparedness, and this book helps elucidate the lessons it has taught or should teach us. But the book also questions some of what may seem to be obvious solutions and may create controversy in doing so. While acknowledging that underground and specially designed above-ground tornado shelters are the only certain places in which most people could survive strong and violent tornadoes, the book asserts that from a statistical standpoint the likelihood of such a tornado hitting a building is too low to justify the expense of installing them. The book also questions whether increasing the lead time for tornado warnings is inherently an important factor in saving lives. The authors tentatively come to the conclusion that more geographically specific warnings with longer lead times might have most value by enabling mobile home residents, and perhaps even those in so-called permanent homes, to travel to safer structures. That, though, raises the concern that increasing numbers of citizens would be caught in tornado-vulnerable vehicles stalled in traffic jams as they tried to flee.

The book is candid in concluding that 2011 may not have been the worst-case scenario of tornado fatalities. While such deadly years and deadly tornadoes are not expected often, the fatality rate in 2011 fell within the expected bounds of the authors' statistical analyses when all factors were considered. Meteorologists, community planners, emergency management officials, and the public will be enlightened by reading the thorough analysis of the tornado hazard explored in this book.

The tornadoes of 2011 shattered the lives of thousands of people and left communities in shambles. Perhaps one source of optimism, though, was offered by the chapter on recovery from tornadoes. In studying past disasters, the authors found that most communities recovered. We can hope that pattern holds true for the ones so devastated in 2011.

Dr. Greg Forbes
Severe Weather Expert
The Weather Channel

THE 2011 TORNADO SEASON IN HISTORICAL PERSPECTIVE

The 2011 tornado year started while champagne corks were still popping to celebrate the new year when an EF-3[1] tornado touched down near Bayette in Attala County, Mississippi, at two minutes past midnight. The first killer tornado of the year occurred on February 1 when an EF-2 tornado struck Franklin County, Tennessee. Through the end of March the tornado season had been relatively benign, with 152 tornadoes and 2 fatalities nationwide, compared with averages of 193 and 27, respectively, through March of the three previous years. Little or no hint had been given that the United States was about to experience a record-breaking month in April and the deadliest tornado day since 1925.

The first tornadoes of April were on the 4th, in a widespread outbreak from Arkansas to Georgia and North Carolina. The tornadoes

---

1. In the last decade the National Weather Service (NWS) switched from using the Fujita Scale of tornado intensity (F-scale) to the Enhanced Fujuita Scale (EF-scale). The EF-scale maintains the 0 to 5 rating for tornadoes, and the numerical categories are intended to be consistent with the F-scale ratings. For more information, see www.spc.noaa.gov/faq/tornado/ef-scale.html.

on this day were almost all weak and no lives were lost. A more serious outbreak occurred in the middle of the month as tornadoes struck on the 14th in Kansas, Oklahoma, and Arkansas, continued on the 15th across Arkansas, Mississippi, and Alabama, and culminated in a deadly string of tornadoes on the 16th centered on North Carolina but ranging from Georgia to Maryland. This three-day onslaught claimed 38 lives, which normally would make it the most significant event of the season. Another multistate outbreak from Oklahoma to Kentucky and Indiana ensued on April 19, followed closely on April 22 by one of the bright spots of the season—from the standpoint of impacts avoided—an EF-4 tornado that hit St. Louis's Lambert Field but resulted in no deaths. The record-setting month concluded with the unprecedented 24-hour outbreak on April 27–28, which we will examine in detail in Chapter 2.

May is normally the peak month for tornadoes, but May 2011 began somewhat slowly, almost as if Mother Nature had to rest after April's fury. The month turned deadly, however, on the 22nd with a tornado in the early afternoon in Minneapolis followed by a devastating EF-5 tornado that tracked across Joplin, Missouri, and resulted in 159 deaths and $3 billion in damage. This was followed just two days later by a violent multistate outbreak across Oklahoma, Kansas, Arkansas, and north Texas. The first day of June brought a killer tornado to Massachusetts, as the deadly season seemed intent on bringing death and destruction across much of the nation. Fortunately June 1 appears to have been a climax to the season as opposed to an ominous beginning of another deadly month, and after June 1 only seven more lives were lost due to tornadoes.

We are economists by training with a personal and research interest in severe weather. We have been conducting research on tornadoes for over a decade now. In February 2011 we published a book examining the impacts of tornadoes on the United States, *Economic and Societal Impacts of Tornadoes*. The special emphasis of our research and that book was on tornado casualties and tornado warnings. No sooner had we received our authors' copies of the book than the record-setting month of April unfolded. We consequently decided that we should revisit tornado lethality in light of the 2011 season. This book represents an attempt to explore why the United States should have experienced such a deadly season after decades of progress in reducing tornado lethality.

1.1. An Overview of the Season

The 2011 tornado season has made headlines for its death toll. We first summarize some aspects of the tornado season and then provide some historical perspective on the season in the next section. In this work we use fatality totals as reported by the Storm Prediction Center (SPC) in Norman, Oklahoma, as of December 2011. Because 2011 is not yet over as we write this, the potential exists for the totals reported here to change before the year is over, and for additional information about fatalities to become available. We certainly hope that the rest of the year will prove uneventful for tornadoes, but readers should understand that some of the figures change slightly when all the data become available. Also note that various sources report different fatality totals for certain tornadoes or the year as a whole. Most often this is because SPC totals include only deaths directly related to a tornado. The damage amounts cited in this book are also subject to revision as repairs take place and insurers close claims.

At press time the 2011 season had featured 59 different killer tornadoes in 14 different states. Figure 1.1 displays the fatality totals for the eight states with the most fatalities in 2011. Alabama has suffered the most fatalities, at 242, followed by Missouri with 159, Tennessee at 33, and Mississippi at 32. To place these totals in perspective, no state had suffered as many as 100 fatalities in a single year since Mississippi in 1971, while 1953 was the last year in which two states had over 100 fatalities. Turning to the individual killer tornadoes, Table 1.1 lists 2011's deadliest twisters, led by the Joplin tornado with 159. The northern Alabama EF-5 tornado and the Tuscaloosa-Birmingham EF-4 tornadoes killed 72 and 64, respectively, giving 2011 three different tornadoes exceeding the 50-fatality mark. No tornado in the United States had killed 50 or more persons since 1971, so three 50-fatality tornadoes in one year after none in 40 years is extremely unusual. The deadly season was primarily the result of violent tornadoes, meaning tornadoes

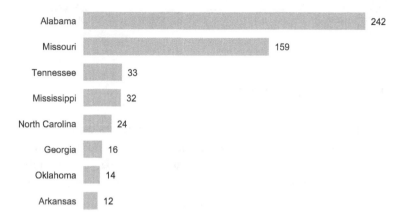

Alabama	242
Missouri	159
Tennessee	33
Mississippi	32
North Carolina	24
Georgia	16
Oklahoma	14
Arkansas	12

FIGURE 1.1. States with the most tornado fatalities, 2011

TABLE 1.1. The Deadliest Tornadoes of 2011

Date	EF-Scale	Location	Deaths
May 22	5	Joplin, Missouri	159
April 27	5	Northern Alabama	72
April 27	4	Tuscaloosa, Birmingham, Alabama	64
April 27	5	DeKalb County, Alabama	23
April 27	5	Mississippi, Alabama	23
April 27	4	St. Clair, Calhoun Counties, Alabama	22
April 27	4	Georgia, Tennessee	21
April 27	4	Fayette, Walker Counties, Alabama	13
April 16	4	Alabama, Georgia	13
April 16	3	North Carolina	12

rated EF-4 or EF-5. Figure 1.2 displays the breakdown of 2011 fatalities by EF-scale category. Over half of 2011 fatalities occurred in EF-5 tornadoes and almost 30% occurred in EF-4 tornadoes. By contrast, five deaths occurred in weak tornadoes (rated EF-0 or EF-1), or less than 1% of the total.

Over the past 25 years, more tornado deaths have occurred in mobile homes than any other location tracked by the NWS, and over the

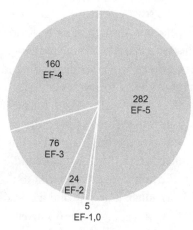

FIGURE 1.2. 2011 tornado fatalities by EF-scale category

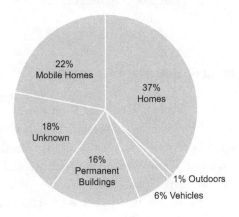

FIGURE 1.3. 2011 fatalities by location

past decade half of all fatalities have occurred in mobile homes. That trend did not continue in 2011, as Figure 1.3 illustrates. More fatalities occurred in permanent homes than in mobile homes, although with the location of almost 100 fatalities still unknown as of this writing, this could change. In fact, no deaths in the Joplin tornado occurred in mobile homes. Over 15% of 2011 fatalities occurred in permanent buildings, such as hospitals, schools, churches, and businesses.

In summarizing the season, we see that 2011 consisted of two historically deadly events: the April 27 super tornado outbreak and the Joplin tornado. The April 27 outbreak was similar in many ways to the April 3, 1974, super outbreak, which actually produced more F4[2] and F5 tornadoes and featured violent tornadoes across more different states (SPC Tornado Archive).[3] The April 27 death toll exceeded that of April 3, 1974, due in large part to the two very deadly Alabama tornadoes. Indeed, April 27, 2011, was the single deadliest day for tornadoes in the United States since the Tri-State Tornado on March 18, 1925. The Joplin tornado was the first tornado to exceed 100 fatalities since the Flint, Michigan, F5 tornado in June 1953, and the deadliest U.S. tornado since Woodward, Oklahoma, in 1947. The occurrence of two extremely deadly outbreaks in the same year is actually not without precedent: 1953 featured a tornado that killed 114 in Waco, Texas, in May followed by the storm system that produced the Flint and Worcester, Massachusetts, tornadoes with a combined 206 deaths on June 8 and 9.

1.2. How Much of an Outlier Was 2011?

How does the 2011 tornado season stack up relative to past U.S. tornado seasons? The 2011 season has resulted in a death toll not seen for decades. Figure 1.4 reports the annual U.S. fatality total since 1900, and confirms this point. Fatalities in 2011 stand at 552, and the 500-fatality threshold has not been eclipsed since 1953, when 515 fatalities occurred. Until the 1950s, the nation experienced a 500-plus-fatality year about once a decade, as the threshold was exceeded five times between 1900 and 1950. The last year the nation experienced more tornado deaths than 2011 was in 1936, with 555, while the deadliest year since 1900 was 1925 with 805 fatalities.

The fatality totals certainly suggest that tornadoes have become less deadly over time. We say *suggest* because fatality totals depend

---

2. Tornadoes occurring prior to 2007 used the Fujita Scale of tornado intensity.

3. http://www.spc.noaa.gov/wcm/

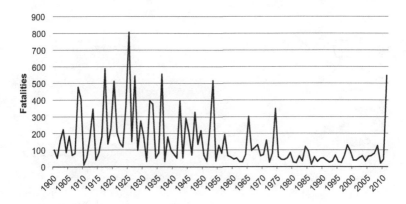

FIGURE 1.4. Annual fatalities (1900–2011)

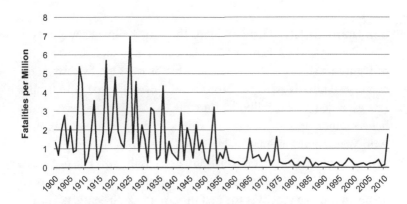

FIGURE 1.5. Fatalities per million (1900–2011)

on the number, strength, and location of tornadoes in a year (e.g., in urban or rural areas). The fatality totals alone do not allow us to say for certain that lethality has decreased, but more thorough analyses do confirm this. The reduction in fatalities before 2011 evident in Figure 1.4 is even more impressive when considering the population growth in the United States during the 20th century. To adjust for population change, Figure 1.5 displays the annual tornado fatality rate, or fatalities per million U.S. residents (based on Census Bureau estimates of the annual population and estimated population as of October 2011). The 2011 U.S. population of 312 million implies that this year's 552 fatalities

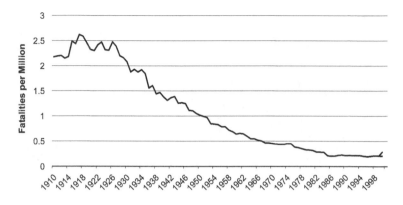

FIGURE 1.6. 20-year moving average—fatalities per million (1900–2011)

translate to 1.75 fatalities per million. The 2011 fatality rate was last exceeded in 1953, when there were 3.21 fatalities per million. Examination of Figure 1.5 shows that 2011's fatality rate is not high by early-20th-century standards (see also Brooks and Doswell 2002). The fatality rate exceeded 3.5 per million, or double the rate in 2011, seven times between 1900 and 1937, with a maximum of 6.95 per million in 1925. A 20-year moving average of fatalities per million (Figure 1.6), which smooths out year-to-year fluctuations, exceeds the 2011 rate until 1935. When considering fatalities per million persons, 2011 would have been about an average year over the first third of the 20th century.

Although annual fatality totals show that the 2011 tornado season was unusual, Figure 1.4 does not necessarily illustrate just how different the 2011 season was from the preceding seasons. The United States has had years with 500 or more fatalities before, but these occurred when far more Americans died in tornadoes in the other, nonpeak years. To illustrate this, we calculated the average and maximum fatalities for the 20 years around several of the deadliest seasons, excluding the peak years themselves so as not to inflate the averages. When available we used the 10 years before and after the season, but for 2011 we use the 20 prior years, 1991 to 2010.

Table 1.2 reports the year's fatality total and the contemporary average and maximum fatalities for nine of the deadliest tornado years. To

TABLE 1.2. Comparing Peak Tornado Fatality Seasons

Year	Fatalities	Recent Average	Recent Maximum	Ratio of Fatalities to Average	Ratio of Fatalities to Recent Maximum
1908	523	286	588	2.56	0.81
1917	588	253	805	2.32	0.73
1925	805	238	543	3.38	1.37
1936	555	183	543	3.04	1.02
1953	515	118	327	4.36	1.57
1965	301	96.0	348	3.14	0.86
1974	348	83.9	301	4.15	1.16
1998	130	54.5	126	2.39	1.03
2011	552	57.6	130	9.58	4.25

measure the extent to which each of these peak years is an outlier relative to fatalities in the surrounding years, Table 1.2 also reports the ratio of current fatalities to the contemporary average and maximum. The ratios demonstrate the truly unprecedented aspect of the 2011 tornado season. The Tri-State Tornado in 1925 produced the deadliest year in the United States with 805 fatalities. Yet at this time the United States averaged 230 fatalities a year (with 1925 excluded from the average). Thus fatalities in 1925 were just more than triple the contemporary average, and only 37% higher than the next-highest toll in the surrounding decades (588 in 1917). By contrast, the 2011 death toll is more than 9 times the recent average of 58 fatalities a year, and more than 4 times greater than the next-deadliest recent year. The 1925 season claimed more lives than 2011, but at a time when 4 times as many Americans died in tornadoes in a typical year. Only in 1953 and 1974 did the death toll exceed the contemporary average by a factor of 4, and 1953 features the second-greatest difference over the next-worst fatality total, with a ratio of 1.57.

Thus the 2011 season looks like the greatest outlier among tornadoes. A natural question then is how it compares to outlier years for other types of extreme weather fatalities. Season fatality totals are

TABLE 1.3. The 2011 Tornado Season versus Other Extreme Weather Seasons

Type and Year	Fatalities	Recent Average	Recent Maximum	Ratio of Fatalities to Average	Ratio of Fatalities to Recent Maximum
Tornadoes 2011	552	57.6	130	9.58	4.25
Lightning 1943	432	241.6	419	1.79	1.03
Floods 1955	302	74.0	188	4.08	1.61
Floods 1972	555	131.0	445	4.24	1.25
Heat 1995	1021	99.3	502	10.3	2.03
Hurricanes 1900	8000	88.5	600	90.4	13.3
Hurricanes 1928	1836	103.7	601	17.7	3.05
Hurricanes 1957	390	35.6	212	11.0	1.84
Hurricanes 1969	256	18.7	122	13.7	2.10
Hurricanes 2005	1853	14.0	64	132.4	29.0

available for hurricanes, floods, lightning, and heat, and Table 1.3 displays several peak years for these hazards. Again, the degree of the outlier year is measured relative to the 20-year average and maximum fatalities over the surrounding 20 years, as described for Table 1.2. The 2011 tornado season is more of an outlier than peak years for lightning and floods, comparable to heat wave fatalities in 1995, but hurricane season totals exhibit more variance than tornadoes. The most extreme outlier of all the types of extreme weather reported in Table 1.3 is the 2005 hurricane season, with 1,853 fatalities, almost all from Hurricane Katrina. The 2005 hurricane season produced a death toll 130 and 30 times greater than the contemporary average and most deadly season. The Galveston hurricane in 1900 resulted in more deaths (8,000) than 2005, but its total is only(!) 90 and 13 times greater than the contemporary average and next-worst season death tolls. Thus the 2011 tornado season does not reflect a unique event across all types of extreme weather. Coincidentally, perhaps, the most extreme outlier years for hurricanes and tornadoes have occurred in the past decade.

Any cloud-to-ground lightning strike can be lethal, and there are millions of such lightning strikes each year in the United States, and so lightning fatalities exhibit little year-to-year variation during a steady decline from around 400 per year in the 1940s to 22 today. By contrast, the United States averages about one hurricane landfall per year,[4] and so the potential lethality risk varies substantially depending on whether a major hurricane or multiple storms make landfall, and whether they strike a highly populated area.

1.3. The Challenge for Researchers

The toll from tornadoes in 2011 should concern researchers on several levels. First, of course, is the tragedy of over 500 lives lost. Each life represents hopes, dreams, and promises that will go unfulfilled, and we can only offer our condolences to the persons injured or to those who lost a loved one in the tornadoes of 2011.

An extreme event like the 2011 tornado season also brings determination to not let this happen again, especially with a hazard like tornadoes, which had seemingly been brought under control. That the United States has not experienced a tornado season as deadly as 2011 for over 50 years makes the death toll more daunting, as if polio had returned to terrorize humanity again. In a world with ever-emerging threats to public health, to have one return from the "Finished" box can seemingly erase decades of progress.

Since Gilbert White in the 1970s natural hazards researchers have emphasized that disasters are not entirely natural but instead a product of actions by humans as well as nature. A disaster or a season like 2011 in which a hazard takes a significantly higher toll than in recent decades immediately raises red flags for researchers. An outlier season could be a product of exceptional meteorological vulnerability, exceptional societal vulnerability, or a combination thereof. Certainly, 2011 was an active tornado season, with 1,488 reported tornadoes through July, a large number of which were the strongest type—those rated EF-4 or EF-5, which historically have accounted for a large proportion of fatalities. Six EF-5 tornadoes and 31 EF-4 tornadoes have occurred

---

4. www.nhc.noaa.gov/pastall.shtml

in 2011, while over the prior 11 years the United States had experienced just 2 EF-5 and 63 EF-4 tornadoes. The April 27–28 outbreak featured over 200 tornadoes across 19 states, a record for a 24-hour period.[5] The possibility also exists that the deadly 2011 season may have resulted from exceptional or worsening societal vulnerability. A number of vulnerabilities or factors that make tornadoes more deadly than might otherwise have been identified include the mobile-home problem, the danger of tornadoes after dark, and tornadoes occurring during the fall and winter months. It may be that the killer tornadoes of 2011 have revealed previously unrecognized vulnerabilities or emerging new sources of vulnerability.

Of particular concern for researchers, forecasters, and the public is that this deadly season may indicate the measures that had made tornadoes less deadly and helped make 500-death seasons and 100-death tornadoes a thing of the past have lost their effectiveness. Overall the tornadoes of 2011 have been well warned-for, and the NWS still has its nationwide network of Doppler weather radars, which improved warning performance and reduced tornado lethality. In 2007 the NWS introduced Storm Based Warnings (SBWs) for tornadoes and other types of severe weather, which warn for a much smaller area directly in the path of a tornado instead of entire counties. Is the 2011 tornado season a sign that warnings have become less effective at preventing casualties, or that SBWs are causing confusion among residents?

The important question, to answer, then, is whether the deadly season is attributable to meteorology, existing societal vulnerabilities, or other factors or vulnerabilities that had previously escaped detection. Or perhaps more accurately, what portion of the death toll is due to each of these factors? This book is an attempt to answer this question, by dissecting the season and analyzing it in light of previously recognized vulnerabilities. The answer to this question provides perspective on the lessons to be drawn from the season. If the deadly 2011 tornado

---

5. The number of reported tornadoes in the United States has been increasing for several decades due to new technology for documenting tornadoes; thus the April 27 outbreak may only have produced the most reported tornadoes in a 24-hour period, but by any measures 2011 was a very active year.

season can be explained based on the number, strength, timing, and paths of the tornadoes, then the lesson of 2011 is that an extreme year for tornadoes can result in a death toll not seen for decades despite the advances in safety. If not, then we must look for new vulnerabilities.

1.4. Outline of the Book

As mentioned above, we recently completed a comprehensive examination of tornado casualties and other societal impacts in *Economic and Societal Impacts of Tornadoes*. We will not repeat all of that analysis here but instead will refer readers interested in this research to that book. The purpose of this book is to explore in depth the tornado deaths of the past year and provide some perspective on the death tolls, which the United States had not experienced in over 50 years. The deadly 2011 season raises a number of concerns for researchers and the public. For instance, have NWS tornado warnings somehow lost their ability to reduce fatalities, or are competing sources of information diluting their effect? What role did population growth in states and communities vulnerable to tornadoes play in the reappearance of 100-fatality twisters and 500-fatality seasons? What lessons can be drawn (or should not be drawn) from the season to help the nation reduce tornado fatalities in the future? The current book thus serves as a complement to our earlier book.

We proceed as follows. The deadly tornadoes of April hit the Southeastern United States hardest, and these states were already recognized as particularly vulnerable to tornadoes. One explanation for at least April's death toll could be the preexisting relative vulnerability of this region. Chapter 2 examines the Southeastern tornado vulnerability in detail to see how the region's demonstrated vulnerabilities played into the April 27 super tornado outbreak. In general, this outbreak does not fit the types of tornado fatalities most prevalent in the region, and so the April 27 death toll does not appear primarily to be a consequence of Southeastern vulnerabilities.

Chapter 3 tries to disentangle the roles that extreme weather and extreme vulnerability played in the 2011 season. We address this question by projecting fatalities in 2011's worst tornadoes using determinants of fatalities and ratios of buildings damaged, the dollar value of property damage, and injuries to fatalities. These different methods all suggest that the number and strength of the season's tornadoes would have produced a death toll in excess of 300 and perhaps close to the season's actual total. We conclude that extreme weather and existing societal vulnerabilities are the most likely culprits for the death toll, not some exceptional societal vulnerability.

In *Economic and Societal Impacts of Tornadoes* we documented the success of NWS efforts over the decades to reduce the lethality of tornadoes, steps like issuing increasingly accurate tornado warnings and installing a nationwide network of Doppler weather radars in the 1990s. Given the NWS's efforts to reduce tornado fatalities, how is a 500-fatality year still possible? Or do our conclusions regarding warnings and Doppler radar have to be revised in light of the 2011 season? Chapter 4 revisits the effectiveness of Doppler radar by adding preliminary records of 2011 tornadoes to our regression analysis. The addition of 2011 tornado data actually has almost no effect on the life-saving effect of Doppler radar. Chapter 4 also discusses the implication of 2011 for the effectiveness of warnings and presents evidence that the overnight and morning tornado outbreak across Alabama, Mississippi, and Tennessee on April 27 contributed to deaths in the afternoon and evening tornadoes, perhaps through power outages that slowed dissemination of watches and warnings.

Chapter 5 considers the question of recovery from tornadoes, which was also touched on in *Economic and Societal Impacts of Tornadoes*. The existing studies of tornadoes suggest that communities tend to recover pretty successfully from tornadoes, and we present some new evidence confirming this based on population and retail sales changes after major tornadoes.

Chapter 6 concludes the book and discusses implications of the deadly season for tornado research. The 2011 season highlighted what we call the permanent home–violent tornado problem. Fatalities in permanent homes occur primarily in violent tornadoes (those rated EF-4 and EF-5), and yet these homes generally provide residents sufficient protection against weaker tornadoes. The protection offered by permanent homes ironically renders construction of tornado shelters and safe rooms cost-ineffective, because violent tornadoes are just too rare. A more cost-effective way to reduce fatalities in long-track, violent tornadoes would be to get residents out of the way or call for limited evacuations for homes in the path. The information about tornadoes necessary to make such evacuations feasible can provide goals for research on tornadoes in the years to come.

2

SOUTHEASTERN VULNERABILITY AND THE APRIL 27–28 TORNADO OUTBREAK

2.1. Introduction: A Record Outbreak in the Wrong Place?

The massive tornado outbreak on April 27–28 featured the most reported tornadoes in a 24-hour period on record in the United States. Tornadoes occurred in 19 states over these two days, but the Southeast bore the worst of the outbreak, with 234 fatalities in Alabama, 32 in Tennessee, 31 in Mississippi, and 15 in Georgia. Earlier in the month, North Carolina experienced 24 fatalities on April 16. Overall, 345 of the 360 fatalities during the record-setting month of April occurred in five Southeastern states.

The April 27 super tornado outbreak would have resulted in a substantial death toll wherever it occurred. But the April death toll may have been a consequence of the vulnerability of Southeastern states to tornadoes. We previously (Simmons and Sutter 2011) constructed a tornado lethality index for 28 states east of the Rocky Mountains based on several regression analyses of fatalities and injuries that include state variables over the years 1950 to 2007. The regression models included variables to control for the Fujita-scale rating of the tornado, the length

TABLE 2.1. Tornado Lethality in Southeastern States

State	Tornado Casualty Index	Rank
Alabama	1.63	6th
Florida	2.56	4th
Georgia	2.65	3rd
Mississippi	0.91	16th
North Carolina	1.34	8th
South Carolina	1.35	7th
Tennessee	1.78	5th

The casualty index is as constructed in Simmons and Sutter (2011) and is estimated for 28 states. Index values range from 0.02 to 3.08, with a higher value indicating that tornadoes result in more casualties in a state in a regression analysis controlling for tornado and path characteristics. The states with the highest index values are Massachusetts and New York.

of the damage path, and demographic and economic characteristics of the counties in the tornado path. The casualty index therefore measures the lethality of tornadoes across states *when controlling for the number, strength, timing, population, and other factors.* Table 2.1 displays the casualty index values and rank for seven Southeastern states: Alabama, Georgia, Mississippi, North Carolina, and Tennessee—the hardest hit states in April 2011—and Florida and South Carolina to fill out the region. Larger values of the index indicate that tornadoes pose greater threat to life and limb, and the index is scaled so that the state where tornadoes are least dangerous (North Dakota) has a value close to zero (0.02). The Southeastern states are among the most vulnerable to tornadoes, with six ranking between 3rd and 8th behind Massachusetts and New York. Georgia is the most vulnerable Southeastern state, followed by Florida and Alabama. Mississippi interestingly has the lowest casualty index value of the Southeastern states, ranking 16th out of 28 states, even though the state has the highest tornado fatality rate per million residents since 1950 of any state. The example of Mississippi illustrates that the casualty index measures the residual or leftover vulnerability after regression analysis controls for many factors that might affect tornado deaths or injuries.

Is the deadliness of April 2011 simply a consequence of the preexisting Southeastern tornado vulnerability? If so, much of the deadly 2011 season would be due to a record-setting outbreak in the most vulnerable part of the country. We explore this possibility in this chapter and begin by examining the pattern of casualties in the Southeast. The region accounts for a disproportionate share of fatalities in mobile homes and in night, morning, fall, winter, and weak and strong (EF categories 1 through 3) tornadoes. We then discuss a regression analysis separating tornadoes in the Southeast and the rest of the United States to explore if the relationship between casualties and tornado and path characteristics differs in these states. Although many factors affect casualties similarly, we find several interesting differences, notably mobile homes and tornado-warning false alarms. We then evaluate the April 27 outbreak in light of the preexisting Southeastern vulnerabilities. April 27 does not fit the mold of the tornadoes where Southeastern states have exhibited their greatest vulnerability, as the outbreak involved multiple violent (EF-4 and EF-5 rated) tornadoes during the afternoon and evening hours in the spring tornado season.

2.2. Patterns of Vulnerability in the Southeast

We analyze tornado casualties from 1950 (the first year in the Storm Prediction Center's [SPC] national tornado archive) through 2010, to focus on patterns *prior to* the 2011 season (see also *Economic and Societal Impacts of Tornadoes*). The statistics we report in this book are based, except where noted, on our calculations using SPC records, and we are responsible for any errors in the calculations. Over these 61 years the United States experienced over 5,000 fatalities and nearly 85,000 injuries, and about 30% of both fatalities and injuries occurred in the seven Southeastern states. We first examine whether fatalities are becoming more concentrated in the Southeast over time. The percentage of fatalities in the Southeast varies substantially from year to year, as typically only several large tornado outbreaks account for half or more of a year's fatalities. In years where these outbreaks occur in

TABLE 2.2. Tornado Fatalities by Decade

Decade	U.S. Fatalities	Southeast Fatalities	Percent in Southeast
1950s	1,415	288	20.4
1960s	942	191	20.3
1970s	973	396	40.7
1980s	520	169	32.5
1990s	567	251	44.3
2000s	603	276	45.8

The 2000s decade includes 2010. States counted in the Southeast for these calculations are Alabama, Florida, Georgia, Mississippi, North Carolina, South Carolina, and Tennessee.

the Southeast, most of the year's fatalities occur in the region. Consequently the annual percentage of fatalities in the Southeast exhibits too much "noise" to readily allow detection of a possible trend. Instead, Table 2.2 reports fatalities by decade since 1950 for the nation and the seven Southeastern states, and shows that the region's share of tornado fatalities has increased over time. About 20% of fatalities occurred in the Southeast in the 1950s and 1960s, and this percentage jumped to 41 in the 1970s. After falling back to 33 in the 1980s, about 45% of fatalities occurred in the Southeast in the 1990s and 2000s.

Population growth offers one possible explanation for the Southeast's increasing share of tornado fatalities. Indeed, several news stories after the April 27 outbreak mentioned population growth in areas prone to tornadoes as possibly responsible for the death tolls. The population of almost every state in the United States has increased since 1950, and so population growth as a cause of the change in the location of fatalities in Table 2.2 must be interpreted as faster growth in the Southeast than in other tornado-prone states. And this has indeed been the case. The population of these seven Southeastern states increased from 21 million in 1950 to 57 million in 2010, while the population of other tornado-prone states increased from 108 million to 176 million during this time.[1] Adjusting for differences in population

---

1. Specifically, the states used for comparison here include all states east of the Rocky Mountains except Maine, New Hampshire, and Vermont.

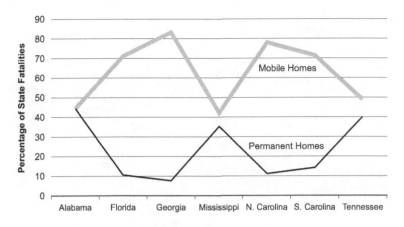

FIGURE 2.1. Tornado fatalities by location in Southeastern states

growth suggests that about 28% of fatalities in the 1950s and 1960s might have occurred in the Southeast if relative populations were as in 2010, or that about one-third of the increase in the percentage of fatalities in the Southeast is due to relative population growth. Thus population growth accounts for some of the time trend apparent in Table 2.2, but other factors must also be at work.

Mobile or manufactured homes are known to be vulnerable to tornadoes, and since 1985, 45% of fatalities have occurred in these homes. The mobile home–tornado problem is largely due to the Southeastern states, and particularly Florida and Georgia. Figure 2.1 reports the percentage of fatalities occurring in mobile and permanent homes in the seven Southeastern states between 1996 and 2010; for comparison, 47% and 34% of fatalities nationally occurred in mobile and permanent homes, respectively, over these years. In the Southeast these percentages are 58 and 28, respectively, and over 70% of fatalities occurred in mobile homes in Florida, Georgia, North Carolina, and South Carolina, although the number of fatalities is much smaller in the Carolinas. The seven Southeastern states together account for 69% of all U.S. mobile-home fatalities over these years. Tennessee and Alabama have the most fatalities among the Southeastern states over these years, with permanent- and mobile-home fatalities more equally balanced in these states. The larger share of fatalities in mobile homes in Florida and

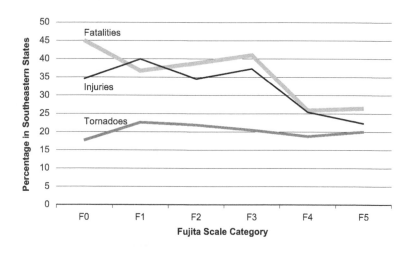

FIGURE 2.2. Southeastern casualties and tornadoes by F-scale category

Georgia masks a greater risk for mobile-home residents in Tennessee or Alabama. The annual fatality rates for mobile-home residents (based on 2000 Census data) are 2.8 in Florida and 3.8 for Georgia compared with 3.9 and 6.4 in Alabama and Tennessee. The exceptional aspect of Florida and Georgia is a near lack of permanent-home deaths; Tennessee and Alabama have experienced high fatality rates in both permanent and mobile homes. The U.S. mobile home–tornado problem is largely a product of the Southeastern states.

All tornadoes are not created equal in terms of societal impacts and particularly lethality. Over 95% of tornadoes result in no fatalities, while the 106 tornadoes between 1950 and 2010 that resulted in 10 or more fatalities accounted for 49% of all deaths over the period. Generally, fatalities are concentrated in the tornadoes rated as strong and violent on the EF-scale. Yet this pattern differs to some extent in the Southeast. Figure 2.2 displays the percentage of U.S. tornadoes, fatalities, and injuries in the Southeast between 1950 and 2010 broken down by EF-scale category. Overall, about 20% of all tornadoes and 30% of fatalities and injuries occur in the Southeast, but casualties in medium-strength tornadoes—those rated EF-1 through EF-3—are more concentrated in the Southeast. In each of these categories, be-

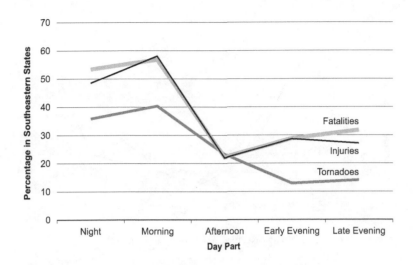

FIGURE 2.3. Tornadoes and casualties in the Southeast by day part

tween 35% and 40% of fatalities and injuries occur in the Southeast. The concentration of casualties in these categories is not due to a concentration of these tornadoes in the Southeast, as no more than 22.5% of tornadoes in any of these categories occur in the region. Southeastern fatalities in tornadoes rated EF-1 to EF-3 tend to occur in mobile homes (see Simmons and Sutter 2009), and so these vulnerabilities appear intertwined.

Tornadoes that occur at night are known to be more dangerous than tornadoes during the day when controlling for EF-scale and other factors (Ashley 2007; Ashley et al. 2008). The lethality of tornadoes after dark also has a Southeastern element. Figure 2.3 reports fatalities and injuries for five different parts of the day (all times local): Night (12:00 a.m.–5:59 a.m.), Morning (6:00 a.m.–11:59 a.m.), Afternoon (12:00 p.m.–3:59 p.m.), Early Evening (4:00 p.m.–7:59 p.m.), and Late Evening (8:00 p.m.–11:59 p.m.). Over half of all night fatalities (53%) and nearly half of all injuries (49%) occur in the Southeast. The concentration of night tornado casualties is due in part to the prevalence of night tornadoes in the Southeast (36%) and because night tornadoes are more dangerous in the Southeast than the rest of the country. Further examination of Figure 2.3 reveals that morning tornadoes and

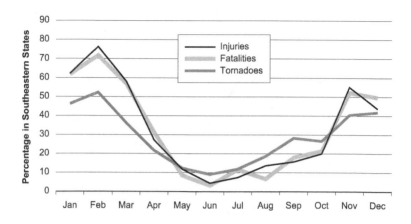

FIGURE 2.4. Tornadoes and casualties in the Southeast by month

casualties are even more concentrated in the Southeast, but morning tornadoes overall are neither particularly frequent nor dangerous (nationally less than 10% of tornadoes and only 5% of fatalities over the period 1950–2010), so we do not emphasize this concentration. Casualties and tornadoes between noon and midnight tend to occur outside of the Southeast, with fewer than 15% of U.S. tornadoes during the early and late evening hours occurring in the region.

The lethality of tornadoes varies substantially over the course of the year, for reasons that are not completely understood (Simmons and Sutter 2011). Figure 2.4 reports the percentages of tornadoes and casualties in the Southeast across the year. The lion's share of casualties during the deadly "off season" fall and winter months occur in Southeastern states. Over half of U.S. fatalities and injuries in each month from November to March, except for December, occur in the Southeast. Even in December the Southeast accounts for 49% and 44% of fatalities and injuries, respectively, while over 70% of February casualties occur in the region. By contrast, fewer than 15% of U.S. fatalities and injuries occur in the Southeast in the summer months from May through August, including less than 3% of June fatalities. Some of the concentration of fall and winter casualties is due to tornado climatology, as over 40% of U.S. tornadoes between November and February occur in the region. Note that the lethality of tornadoes in the Southeast relative to the rest of the nation appears to vary as well across the year. While the region typically accounts for a higher proportion of casualties from tornadoes, the opposite is generally true between May and August. Thus, although tornadoes are overall more dangerous in the Southeast than other regions, everything else equal—which is what the state casualty index values in Table 2.2 tell us—tornadoes appear to be particularly dangerous in the Southeast between November and April.

2.3. Comparing Southeastern Vulnerabilities to Other Regions

We now further explore the Southeastern vulnerability through regression analysis of fatalities and injuries. Regression analysis allows researchers to determine how each factor affects casualties when combined with the other determinants of casualties, or the effect that a change in one factor will have on casualties holding other variables constant. We have typically estimated casualty models by using tornadoes from across the nation in one regression. Here we estimate separate models for the Southeastern states and for the rest of the nation, which allows us to see if the direction and magnitude of several variables differ between the Southeast and the rest of the nation. Such differences could provide further evidence on the Southeastern vulnerability.

The full regression specifications are available as part of this book's supplemental online material. The EF-scale rating of a tornado is the most important driver of fatalities, but the effect of EF-scale rating is statistically indistinguishable except for EF-5 tornadoes. We immediately see substantial differences for tornado-warning variables. Simmons and Sutter (2009) recently found that a higher, local false-alarm rate (FAR) increases casualties. This result is due to tornadoes outside of the Southeast, as a higher FAR significantly reduces fatalities in the Southeast. The effects are substantial in each region, with a one standard deviation increase in the FAR decreasing expected fatalities by 11% in the Southeast and increasing expected fatalities 32% in the rest of the nation. A higher FAR increases expected injuries in each part

TORNADO CASUALTIES REGRESSIONS

At various points throughout this book we will discuss results from a multiple regression analysis of tornado casualties. The regression analysis considers how all of the various factors that might affect the lethality of tornadoes, including storm characteristics, storm path characteristics, and tornado warnings, jointly affect fatalities or injuries. Regression analysis is needed to separate out the effect of the different variables that change when comparing different events. For instance, we might expect that mobile homes and income are both important determinants of tornado fatalities, and we might find that fatalities are higher for tornado paths with lower incomes and more mobile homes. But we might also find that lower income paths tend to have more mobile homes, and thus univariate comparisons would be unable to identify how a change in income or mobile homes alone affects fatalities. Multiple regression analysis allows us to parse out these effects.

Regression analysis is also important because most actions society might take to reduce casualties are only expected to reduce casualties *ceteris paribus*, or everything else equal. Tornado warnings could be expected to reduce casualties, but they are not the only factor that affects casualties. An analysis that simply compares fatalities in warned tornadoes with unwarned tornadoes could easily find higher casualties for the warned tornadoes, because other factors increasing casualties might correlate with warnings. Thus, although the regression analysis is somewhat complicated, it was an important component of our research effort.

We will draw on some new regression analysis but will not present the results or offer detailed definitions of the variables or econometric methods employed. Readers interested in this level of analysis can turn to *Economic and Societal Impacts of Tornadoes* for a detailed discussion of methods employed and this book's online supplemental material for the new regression specifications referenced here. Here we will only provide an overview of the data used in the analysis of the 2011 tornado season. Most of the factors we discuss in the text are based on variables defined in a relatively straightforward manner; for instance, income will be inflation-adjusted median family income in storm path.

The analysis includes three groups of variables, which measure tornado characteristics, path characteristics, and tornado-warning variables. The tornado characteristics include the EF-scale rating of the tornado, the damage path length, and the timing of the tornado. The path characteristics are economic and demographic character-istics that might affect casualties, such as population density, the proportion of mobile homes in the housing stock, and income. Warn-ing variables include both warnings issued on a tornado and the false-alarm ratio. All of the tornado characteristic variables employed in the regression analysis are constructed from the tornado records in the SPC archive, while the economic and demographic variables use Census data for the counties listed as in the path by the SPC archive. Census data were used from 2000 and earlier, with annual values established by linear interpolation from the decadal values. Census data from 2010 were not yet available for most states at the time of this analysis, and thus the economic and demographic variables for tornadoes since 2000 are taken from the American Com-munity Survey.[2] The path variables for tornadoes striking more than one county average the values for each county. The warning variables are constructed from the National Weather Service (NWS) national warning verification records.

---

2. www.census.gov/acs/www/

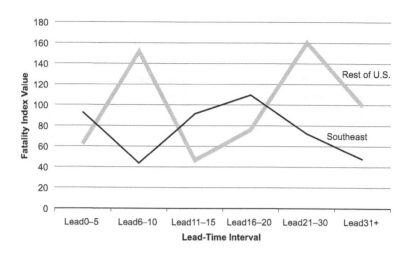

FIGURE 2.5. Fatality index by lead-time interval

of the country, with similar point estimates, but only attains statistical significance in states outside of the Southeast. Figure 2.5 explores the effect of tornado warnings with various lead times on fatalities and injuries. Warnings generally reduce expected casualties in both the Southeast and the rest of the nation. The greatest reduction in fatalities in the Southeast is observed for a 6- to 10-minute lead time (56%) relative to no warning, followed closely by a 53% reduction for lead times in excess of 30 minutes, while the greatest reduction outside of the Southeast is for an 11- to 15-minute warning (53%). Previous research has documented that longer lead times are associated with more fatalities; we have a statistically significant increase in fatalities only outside of the Southeast, and in the 6- to 10-minute and 21- to 30-minute intervals. Injuries show consistent reductions for warnings in each lead-time category in each part of the country, although the effect is not statistically significant in the Southeast. The greatest reductions are in the 6- to 10-minute interval in the Southeast (48%) and for lead times over 30 minutes in the rest of the United States (53%).

Timing matters significantly for tornado casualties. Figure 2.6 displays an index of tornado lethality for times of the day for fatalities and injuries in the Southeast and the rest of the nation. The index is

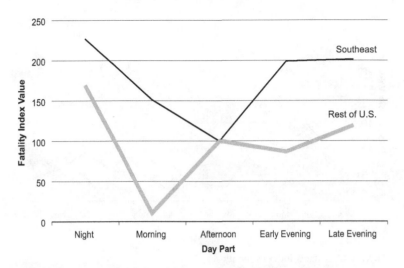

FIGURE 2.6. Fatality index by time of day

constructed so that a tornado during the afternoon hours has a value of 100, and lethality at other times relative to the afternoon is given by the value divided by 100. In both the Southeast and the rest of the United States, tornadoes at night are more deadly than at other times of the day, although the magnitude of the night lethality is greater in the Southeast. Some differences in the timing pattern are apparent, though. Tornadoes during the evening hours in the Southeast are almost as deadly as nighttime tornadoes and twice as deadly as afternoon tornadoes, while evening tornadoes in the rest of the United States are about as deadly as afternoon tornadoes. This difference in lethality was not readily apparent in Figure 2.3, which presents the proportion of casualties and tornadoes by day part in the Southeast. But tornadoes after 4 p.m. are almost as deadly as night tornadoes in the Southeast. And the reason for the concentration of morning fatalities in the Southeast becomes readily apparent in Figure 2.6. Morning tornadoes are moderately deadly in the Southeast, but rarely lethal in the rest of the United States, where afternoon and night tornadoes are 10 and 17 times more deadly than morning tornadoes. For injuries the pattern across the day is similar in both parts of the country, with injuries lowest in afternoon tornadoes and highest in night tornadoes.

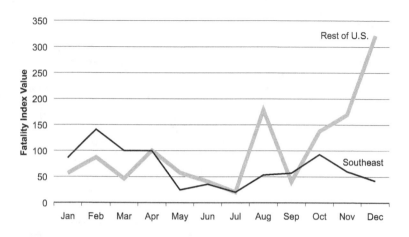

FIGURE 2.7. Fatality index by month

Injuries actually display little variation across the day in the Southeast, with a maximal difference of 35% between night and afternoon (which is the only statistically significant difference). Injuries are 50% higher in the rest of the United States in night and morning tornadoes relative to the afternoon.

Figure 2.7 reports an index showing the effect of month on expected casualties, controlling for other factors. The index is constructed in each case so April has a value of 100. The figure reveals that the spring months of February, March, and April are the most lethal in the Southeast and the fall months of October, November, and December are the most deadly in the rest of the United States. The magnitude of variation in fatalities and injuries across the year is substantial both in the Southeast and the rest of the United States, with expected fatalities 7 times greater in February than July in the Southeast and 16 times greater in December than July in the rest of the United States. The same pattern across the year is apparent for injuries in each region.

Many of the included demographic variables have similar effects across the two regions, but several notable differences emerge. Of most interest is the proportion of mobile homes in the county housing stock, which increases fatalities and is statistically significant for tornadoes in the Southeast but which is insignificant outside of these states. Previously this variable has been found to be a consistent determinant of fatalities (Simmons and Sutter 2011), but here we see that the mobile-home problem may be largely regional, as Figure 2.1 suggests. Mobile homes increase expected injuries in each part of the country, although the effect is smaller than for fatalities in the Southeast, and the injuries results are only on the margin of statistical significance. There also appears to be a more pronounced demographic pattern of fatalities in the Southeast, as the poverty rate and the Gini coefficient[3] increase expected fatalities in the Southeast but not in the rest of the nation. This suggests that inequality and poverty may play a role in fatalities in the Southeast, although these patterns do not carry over to injuries (both variables are insignificant in both areas). A higher minority population significantly decreases fatalities in the Southeast but not the rest of the nation.

2.4. Assessing the Record Outbreak

We can now assess the April 27 tornado outbreak relative to the pre-existing Southeastern vulnerabilities. The outbreak does not actually accord with the traditional Southeastern vulnerabilities. The vast majority of fatalities on April 27 occurred in violent tornadoes. Violent tornadoes accounted for 85% of fatalities on April 27, as opposed to 50% of Southeastern fatalities historically. A high proportion of fatalities in violent tornadoes will be typical of any super tornado outbreak, simply because violent tornadoes rarely occur outside of major outbreaks. The Southeast region had not experienced an EF-5 tornado since 1998, so necessarily zero percent of the region's fatalities between 1999 and 2010 occurred in EF-5 tornadoes. Nonetheless, fatalities on April 27 were primarily due to EF-4 and EF-5 tornadoes as opposed to the EF-1- through EF-3-rated tornadoes where the Southeast has exhibited vulnerability.

The outbreak also occurred during the spring and during the afternoon and early evening hours. Southeastern vulnerability has been concentrated in the late fall and winter (or early spring) months from

---

3. A Gini coefficient is a measure of income inequality.

November to March. And 92% of fatalities on April 27 occurred in tornadoes that touched down between 1:30 p.m. and 8 p.m., while night (and morning) tornadoes were where a disproportional percentage of fatalities occurred in the Southeast. April 27 was a traditional spring afternoon and evening event, not the after dark, off-season event so common in the Southeast.

Mobile homes were not a major part of the vulnerability. Figure 2.8 displays fatalities by location in the April 27 outbreak, as reported by the SPC at of the end of September. Over one quarter of the fatalities were still reported as location unknown as of this writing, and eventual determination of the location of these fatalities may significantly alter the patterns observed here. Figure 2.8 displays the location only of deaths with a known location. Mobile and permanent homes account for 35% and 52%, respectively, of April 27 fatalities, just about

THE NORTH CAROLINA TORNADO OUTBREAK: A MORE TRADITIONAL SOUTHEASTER KILLER EVENT

While the April 27 super tornado outbreak did not really fit the picture of the types of tornado fatalities that occur disproportionally in the Southeast, the North Carolina outbreak 11 days earlier is a more traditional Southeastern killer tornado event. Thirty-one tornadoes occurred in North Carolina on April 16, resulting in 24 deaths, over 440 injuries, and nearly $400 million in estimated property damage. The fatalities on April 16 occurred in strong as opposed to violent tornadoes, with 20 in EF-3-rated county segments and 4 in EF-2-rated segments. More deaths occurred in mobile homes (14) than in permanent homes (9) in this outbreak (14 vs. 9). The tornadoes occurred on a Saturday, and tornadoes are more deadly on weekends than during the week. Regression analysis on the Southeastern states shows that expected fatalities are 37% higher for a weekend tornado.

North Carolina had experienced 100 tornado fatalities over the years 1950–2010, and so 24 fatalities on one day might seem like

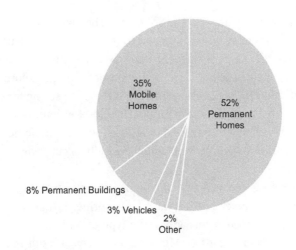

FIGURE 2.8. April 27 fatalities by location

a huge addition to this state's total. Tornado fatalities tend to cluster in large outbreaks. North Carolina had 35 killer tornadoes over the previous six decades, and seven of these tornadoes and 42 deaths occurred on one day, March 28, 1984. So April 16 was not the deadliest tornado day in the state's history. And Alabama suffered 235 fatalities on April 27, or 64% of the state's fatalities over the previous six decades.

The infrequency of super outbreaks should caution not to infer too much from years or even decades in a state with few tornado deaths. A state might appear to have a low fatality rate but still face significant risk if fatalities are evaluated over a long enough period to include some super outbreaks. The infrequency of super outbreaks can also render evaluation of the effectiveness of measures to reduce fatalities problematic. One could look at the 25 years after the March 28, 1984, outbreak and conclude that North Carolina tornado lethality had been substantially reduced. The infrequency of super outbreaks indicates the need to assemble years or decades of records to adequately evaluate measures to reduce fatalities.

the reverse of the recent pattern in the Southeast. The higher proportion of permanent-home fatalities is related to the number of violent tornadoes in this outbreak, as permanent-home fatalities tend to occur primarily in violent tornadoes. Single-family homes typically afford reasonable protection for residents against weak and even strong tornadoes, provided residents shelter in an interior closet or bathroom. Large numbers of permanent-home fatalities usually occur in homes that have been completely destroyed, which typically occurs only in a violent tornado.

Overall, the April 27 outbreak does not fit the pattern of exceptional Southeastern tornado vulnerability. It was a large, spring, daylight outbreak with numerous violent tornadoes, the type of outbreak more often associated with the Plains or Midwest, but likely to result in a high death toll wherever it occurred.

2.5. Conclusion

The vast majority of fatalities in April 2011 tornadoes occurred in five Southeastern states, a region of the country where historically a disproportionate share of fatalities has occurred. We have explored Southeastern tornado fatalities here to evaluate to what extent the deadly month of April was a product of the preexisting Southeastern vulnerability. To the extent this were so, we could conclude that the deadly 2011 season was in part a consequence of tornado outbreaks in the most vulnerable area of the country. Southeastern states account for a disproportionally large share of deaths in mobile homes, nighttime tornadoes, fall and winter tornadoes, and weak and strong tornadoes. The historic April 27 outbreak did not fit this pattern, as the outbreak featured violent tornadoes striking in the afternoon and early evening hours during the traditional spring tornado season. Moreover, a majority of fatalities occurred in permanent homes. April 27 did not consist of the types of tornadoes that have proven relatively deadly in the Southeast but more benign elsewhere.

This is not to say that Southeastern vulnerability did not contribute to the death tolls. Evidence suggests that the types of tornadoes that claimed so many lives on April 27 are more lethal in the Southeast than elsewhere. Figure 2.7, for instance, shows that springtime tornadoes remain particularly lethal in the Southeast through April before declining substantially in May. Fatalities per evening and late evening tornado are higher in the Southeast than the rest of the nation at 0.22 and 0.26 compared with 0.08 and 0.09. And a regression analysis that uses all tornadoes nationally but includes a variable to control for the Southeast indicates that expected fatalities and injuries are 31% and 47% higher in the region, controlling for tornado characteristics, path characteristics, and warnings. Thus all tornadoes seem to be at least somewhat more dangerous in the Southeast, for reasons that remain unclear. In April 2011, 345 fatalities occurred in the Southeast. This analysis suggests that an extra 70 fatalities might have occurred in the Southeast in April due to the region's elevated vulnerability.

EXTREME VULNERABILITY VERSUS EXTREME WEATHER IN THE 2011 SEASON

3

Natural hazards researchers have long recognized that disasters are not merely acts of God but rather the joint product of natural events and human actions, like building on flood plains or in coastal areas vulnerable to hurricane storm surge (Mileti 1999). From this perspective, an extreme year for tornado fatalities like 2011 would be a product of either extreme weather (many long-track strong or violent tornadoes) or extreme societal vulnerability (perhaps a previously unrecognized form of vulnerability), or both in combination. Both the weather and vulnerability have surface plausibility as explanations. April 2011 featured a record number of tornadoes, and the April 27 outbreak (which accounted for more than half of 2011 fatalities) had a record number of tornadoes for a 24-hour period. A total of six EF-5 tornadoes have occurred in 2011, compared to just two EF-5 tornadoes nationally over the prior decade. In terms of vulnerability, most of the April deaths occurred in the Southeastern United States, which is known to be more susceptible to tornado fatalities (Boruff et al. 2003; Ashley 2007; Simmons and Sutter 2011). And the outbreak on the afternoon and evening of April 27 had been preceded by strong storms that morning that had knocked out power across much of northern Alabama and Mississippi and could have disrupted the normal warning dissemination process.

We apply several types of evidence to help us discriminate between extreme weather versus extreme vulnerability in the 2011 tornado season. One type of analysis we use is a regression analysis of tornado fatalities to generate out-of-sample predictions for some of the 2011 killer tornadoes. We ask if the tornadoes of 2011 produced fatalities similarly to earlier tornadoes; an affirmative answer suggests that vulnerability was unchanged, and that the number, strength, and paths of this year's tornadoes were responsible for the death toll. We find generally that at least one model predicts fatalities relatively well: For 13 of the 33 tornadoes examined, the observed fatality total is within the confidence interval for predicted fatalities. Fatalities are under- and overpredicted for some individual tornadoes, and we discuss the patterns that might be apparent in the errors. Overall, our results suggest that extreme weather largely accounts for the deadly 2011 season, not exceptional vulnerability.

3.1. Warning Regression Model

We estimate and apply a multiple regression model of tornadoes in our analysis. The analysis here updates a model we used to estimate possible death tolls in worst-case, long-track EF-5 tornadoes (Simmons and Sutter 2011). The model includes a wide range of variables, including characteristics of the path counties (e.g., population density, proportion of mobile homes in the housing stock) and characteristics of the tornado itself. We have used regression analysis to examine determinants of fatalities, with special emphasis on the effect of warnings, false alarms, and Doppler radar, and have not sought to predict fatalities out-of-sample. The economic and demographic variables used in the regression analysis use "path counties," even though tornadoes are quite small relative to the land area of most counties. The county-based variables perform well in the regression analysis, which is based on over 25,000 tornadoes and in which some of the inaccuracy resulting from county-level variables cancels out. Thus we do not expect a model to predict fatalities perfectly by any stretch of the imagination. It does, however, provide a way to begin assessing the role of societal vulnerability in the 2011 season.

Our regression model does not include tornado-warning variables, which have been demonstrated to be significant determinants of fatalities (Simmons and Sutter 2008, 2009). In October 2007 the National Weather Service (NWS) began issuing Storm Based Warnings (SBWs) for tornadoes, which warn for a polygon around the circulation, whereas previously warnings were issued for entire counties. A single county will now often have several segments of path with warnings, rendering county-warning-based variables previously used in our analysis inconsistent with the more recent warning records. Research to date has not settled on a way to define warning variables consistently over the entire period. Our goal here is to use an existing regression model to predict fatalities as opposed to innovating in modeling warnings and false alarms. If we innovated with warning variables in this analysis, we would be unsure if the out-of-sample predictions were modified by the method used to model SBW. Also, roughly about four years of tornadoes with SBWs represent too short of a data sample to estimate a robust model of fatalities. Consequently, we use a longer sample but with no warning variables. The model used here is similar to the one we used in *Economic and Societal Impacts of Tornadoes* to estimate fatalities in a worst-case tornado scenario.

3.2. Do Fatalities Regressions Anticipate the 2011 Death Tolls?

We now assess the contribution of extreme weather and vulnerability via out-of-sample predictions with fatality models. We estimated four separate models, based on tornadoes from 1950–2010 and 1986–2010, respectively, and with and without state dummy variables. The full models are available in the supplemental online materials. A list of 2011 killer tornadoes was obtained from the Storm Prediction Center (SPC) website, and path survey data (primarily damage path length) from various NWS Weather Forecast Office (WFO) websites. We limited attention to the major killer tornadoes and could obtain path lengths

TABLE 3.1. Projected Fatalities for 2011 Tornadoes from Fatalities Regressions

Model	Point Estimates	Lower Bounds	Upper Bounds
1986–2010, State Effects	321.4	127.6	809.3
1986–2010, No State Effects	261.8	103.6	662.1
1950–2010, State Effects	113.5	13.7	943.1
1950–2010, No State Effects	140.2	14.6	1349.9

Projections sum the point estimates and 95% confidence intervals for 33 of 2010's killer tornadoes, which together account for 505 tornadoes.

for 33 of the 59 killer tornadoes of 2011 (through August), tornadoes that collectively accounted for 504 of the year's 552 fatalities. For each tornado we used 2010 Census data for the demographic and economic variables averaged for the counties in the tornado path as reported in the NWS WFO surveys.[1] If 2010 Census data were not yet available, we turned to data from the American Community Survey, and then, if necessary, the 2000 Census for demographic variables.

Table 3.1 reports total fatalities for the 33 tornadoes included in this analysis and the sum of predicted fatalities for each of the four fatality models we estimated. The table also includes the sum of the lower and upper bounds of the 95% confidence interval around the point estimate of fatalities for each of the 33 tornadoes. The observed fatality total of 504 is contained within the 95% confidence interval for each of the four models, although in some cases this is due to a lack of precision, meaning an extremely wide confidence interval. The models estimated with 1986–2010 tornadoes most closely approximate the observed fatality total. This is not a huge surprise, as these models include a richer set of control variables and may better capture changing tornado lethality over time. The 1950–2010 estimates include a significant downward trend in fatalities, which results in lower predicted deaths in 2011. The

---

1. In many cases the counties struck identified by the WFO differed from those reported by the SPC. We used the time reported by the SPC, even though these differed in some cases from the times reported by the WFO, but given the day parts employed, the modest differences in time would result in very few tornadoes being coded differently.

TABLE 3.2. Projected Fatalities in Selected 2011 Tornadoes

State	Date	EF-Scale	Fatalities	Lower Bound	Point Estimate	Upper Bound
North Carolina	4/16	3	12	0.2	0.5	1.2
Mississippi	4/27	5	3	3.6	9.0	22.5
Alabama	4/27	4	6	1.2	3.1	7.9
Alabama (northern Alabama EF-5)	4/27	5	72	56.8	143.0	360.2
Mississippi	4/27	5	16	4.1	10.2	25.8
Alabama	4/27	4	11	1.0	2.4	6.1
Alabama	4/27	4	13	4.9	12.3	31.0
Alabama (Tuscaloosa)	4/27	4	64	3.5	8.7	22.0
Alabama	4/27	4	1	1.0	2.6	6.6
Mississippi	4/27	4	7	0.6	1.6	3.9
Alabama (DeKalb County)	4/27	5	26	11.2	28.2	71.0
Alabama	4/27	3	22	1.0	2.4	6.0
Georgia (Catoosa)	4/27	4	8	1.3	3.2	8.1
Tennessee	4/27	4	13	1.5	3.7	9.2
Alabama	4/27	4	7	1.5	3.9	9.8
Tennessee	4/27	4	4	2.2	5.6	14.0
Missouri (Joplin)	5/22	5	159	8.3	21.0	52.8
Oklahoma (El Reno)	5/24	5	9	17.6	44.4	111.9

1986–2010 model with state fixed effects also better predicts the observed fatalities than the model without state effects, which suggests that the pattern of tornado lethality across states explains some of the observed differences in fatalities.

Table 3.2 reports predicted fatalities in the most significant tornadoes of 2011, namely those that killed 10 or more persons or EF-4 or EF-5 killer tornadoes. As can be seen, all but three of these tornadoes were part of the April 27 outbreak. We report the point estimates and 95% confidence intervals for the model that most closely predicted observed fatalities, the 1986–2010 model with state effects. Examination reveals that the model underestimated fatalities for the two included EF-3 tornadoes, while the largest predicted totals were for EF-5 tornadoes. This reflects the strong dependence of fatalities in our regression models on the EF-scale rating of the tornado. Among the six EF-5 tornadoes, the model underestimated fatalities in the Joplin tornado, projecting a total of 21 compared to the actual count of 159. This might not seem surprising in one sense, as the Joplin death toll exceeded by almost a factor of 5 the greatest death toll of any tornado in the 1986–2010 sample. But the model did anticipate the potential for a death toll far in excess of recent history, with a predicted fatality count of 143 for the northern Alabama EF-5 tornadoes, and this projection is very close to the death toll in Joplin. Fatalities in the northern Alabama tornado were only about half of the predicted total, at 72. In addition, the model projected a death toll of 44 for point estimate of fatalities in the May 24 El Reno, Oklahoma, EF-5 tornado, a projection exceeding the death toll in the 1999 Bridgecreek–Moore, Oklahoma, EF-5 tornado (36). The difference in estimated fatalities among EF-5 tornadoes stems largely from the path lengths, as the Joplin tornado had a path length of just over 20 miles, compared to 75 for the El Reno tornado and 130 miles for the northern Alabama tornado. The model's point estimate of 28 deaths in the DeKalb County, Alabama, tornado is very close to the actual total of 26. Overall, the model's projected death toll exceeded the observed total in four of six EF-5 tornadoes. The model failed to predict the large death tolls in EF-4 tornadoes, most notably the Tuscaloosa tornado, with a projected 9 deaths compared with the actual count of 66, and the upper bound of the 95% confidence interval was only 22. Interestingly, if the Tuscaloosa tornado had been rated EF-5, the point estimate of expected fatalities would rise to 92, which is reasonably close to the observed total of 63. The Enhanced Fujita Scale offers ultimately a six-point classification of tornado damage, with tornadoes rated based on the worst damage along the path. There will naturally be a fair amount of variation for tornadoes within categories. Wind speeds in Tuscaloosa were estimated by researchers to be 190 mph, which is a high-end EF-4 tornado. The tornado appears to have maintained EF-4 strength throughout much of its 80-mile path length. Our analysis suggests that this tornado produced a fatality total more consistent with an EF-5 tornado. Overall, the 95% confidence interval includes the actual fatality total for 9 of the 18 tornadoes examined in Table 3.2. Of the cases where the actual total falls outside of the confidence interval, actual fatalities exceeded the upper bound in seven cases, while two tornadoes were less deadly than projected.

The casualties models we have developed were designed to assess determinants of casualties, with specific reference to forecast-related variables like warnings and false alarms. The models used county-level data for storm path variables, while the area of a tornado track is very small relative to most counties. The county-level variables have performed well in casualties analysis, but they limit the predictive power of the model for specific tornadoes. While not intended for out-of-sample forecasting, the fatalities regressions seem to provide a reasonable first take on the role of extreme vulnerability versus extreme weather in the 2011 tornado season. It is not surprising that the model fails to predict fatalities for some of the worst tornadoes. Of more interest is whether extrapolation based on recent patterns of fatalities could have anticipated death tolls in excess of the most deadly recent tornado (36) in some of the 2011 killer tornadoes. And the answer here is a clear "yes"; a 100-fatality tornado could have been expected in 2011, as the northern Alabama EF-5 tornado had a projected fatality total very close to that observed in Joplin, and the upper bound of the 95% confidence interval for the El Reno tornado exceeds 100. Furthermore, the pattern of lethality in 1986–2010 suggested that 2011 would be a very bad year, as the lower bound of the confidence intervals of projected fatalities for the 33 tornadoes examined here exceeded 100, and the point estimates suggested the worst season since 1974.

3.3. Projecting Fatalities Using Damage and Injuries

Fatalities regression analysis provides one perspective on whether the deadly 2011 season was a result of the number, strength, and paths of the year's tornadoes. Given the inherent limitations of the analysis, most significantly that the tornado paths were described only at the county level, the out-of-sample forecasts provide a perspective but hardly the definitive perspective on casualties. Other perspectives can be obtained by using other tornado impacts that correlate with fatalities. Specifically, in this section we consider the number of buildings damaged, the dollar amount of property damage, and total injuries. The pattern of lethality in recent tornadoes is represented by fatalities per building or home struck, fatalities per million dollars of damage, and injuries per fatality, respectively. These rates are then used to project deaths in 2011 tornadoes. Damage- or injuries-based analysis can certainly control much more readily for a tornado that tracked through the urbanized portion of a county, like the Tuscaloosa and Joplin tornadoes. A limitation is that the quality of data is lower for damage or injuries.

We begin by considering casualties per building damaged or destroyed. More than four out of every five fatalities in 2011 (81%) occurred in tornadoes rated EF-4 or EF-5 on the EF-Scale, and thus we will focus on buildings struck by (and fatalities in buildings in other recent) violent tornadoes in our comparison analysis. Our primary source for information on numbers of buildings damaged is the online Storm Events database maintained by the National Climatic Data Center (NCDC) in Asheville, North Carolina. The NCDC database reports tornado events by county, and each county along the path of a long-tracked tornado receives its own EF-scale rating. Thus we collect data only on counties along the path of a tornado where EF-4 and EF-5 damage occurred, although in some cases totals are available only for the entire tornado path.

We compiled records on buildings damaged in tornadoes over the past 15 years. We focus on EF-4 and EF-5 tornadoes here as they accounted for over 80% of fatalities in 2011. Due to the small number of EF-5 tornadoes, we use available records for all of these tornadoes since 1996, a total of six tornadoes. For EF-4 tornadoes we use all tornadoes from 2002 to 2010. The sources of data are NWS WFO Web pages, which often archive information on notable weather events in their forecast and warning area, and the NCDC Storm Events database. A narrative or Web page is useful for our purposes here if it mentions the number of homes and/or buildings damaged or destroyed in the tornado. We combine all types of buildings (e.g., homes, businesses, mo-

TABLE 3.3. Fatalities per Building before 2011

EF-Scale	Fatalities In Buildings	Buildings	Fatalities per Building
4	39	7,746	0.00504
5	110	8,425	0.01306

TABLE 3.4. Projected Fatalities for 2011 Tornadoes

Tornado	EF-Scale	Buildings Damaged	Deaths In Buildings	Fatalities per Building	Projected Fatalities
Joplin	5	7,500	142	0.01893	97.9
Tuscaloosa	4	10,997	63	0.00564	55.4
DeKalb County	5	963	35	0.03635	12.6
Northern Alabama	5	2,444	61	0.02456	31.9
Alabama EF-4s	4	18,472	127	0.00688	93.0
Alabama EF-5s	5	3,407	96	0.02818	44.5

bile homes) into one total of buildings damaged. We also combine all levels of damage (destroyed, major damage, minor damage, affected) into a total number of buildings in the tornado path. We do this in part because not all narratives distinguish between damage levels, and restricting attention to only tornadoes with common damage levels would result in a very small sample of usable records. Also damage levels may be somewhat subjective, so what comprises "major damage" might differ across tornadoes even when the common term is used.

Table 3.3 reports the results for recent violent tornadoes. We also tabulate the number of fatalities in buildings (not total fatalities), for the included tornadoes allow calculation of a fatality rate. Our sample of included tornadoes produces roughly equal numbers of buildings damaged in EF-4 and EF-5 tornadoes, while 39 deaths in buildings occurred in the included EF-4 tornadoes as compared with 110 for EF-5 tornadoes. As a result, fatalities-per-building struck are about two and a half times greater for an EF-5 tornado, reflecting their greater lethality.

Table 3.4 presents totals for some of the notable 2011 killer tornadoes for which we could find the number of damaged buildings. Note that the totals for Alabama are for the county on April 27 and combine all buildings damaged if multiple tornadoes struck a county. The dollar value of damage in the second EF-3 or weaker tornado was in each case just a fraction of the dollar value of the violent tornado (when damage estimates were available for both tornadoes), and thus the number of buildings damaged for the entire county is likely close to the total for the violent tornado. The killer tornadoes of 2011 were in each case more deadly than would be projected based on past fatalities, but the number of buildings damaged does project substantial fatality totals for several tornadoes. For instance, with the historical fatality rate for EF-4 tornadoes, nearly 11,000 buildings damaged or destroyed in the Tuscaloosa–Birmingham tornado resulted in a projected 55 fatalities in buildings, compared with the 62 actually observed. This result is of note because the fatalities model substantially underprojected fatalities for this tornado, suggesting a particular vulnerability. The estimated 7,500 buildings damaged or destroyed in Joplin project to 98 fatalities in buildings, compared with the observed total of 142. Projected fatalities in buildings were only about half of the observed total for the northern Alabama EF-5 tornado. The projections here include only tornadoes for which numerous fatalities occurred, whereas the prior tornadoes include tornadoes for which no fatalities occurred. As a result, the projections based on historical rates do not include zero-fatality 2011 tornadoes in which some fatalities would be projected to have occurred. We are not primarily interested in predicting the total number of deaths for the season but rather seeing if death tolls far in excess of recent U.S. experience could have been anticipated in a buildings-level analysis.

A similar type of analysis can be undertaken using the dollar value of property damage caused by a tornado. The dollar value of property damage provides a measure of the overall societal impact of a tornado, or at least the interaction with the built environment, but as other researchers have noted previously, damage estimates for natural disasters are notoriously inaccurate (Gall et al. 2009; Downton and Pielke 2005). While deaths can occur outdoors or in vehicles (between 1996 and 2010, 4.9% and 9.5% of fatalities have occurred, respectively, in each of these locations), most people do shelter in buildings during tornadoes, and so damage to the built environment represents a way to control for the potential threat to life. Damage amounts will provide a better measure of when a tornado strikes a populated area of a county, relative to county variables and storm paths in regression analysis. Of course damage is not necessarily a perfect control for numbers of people at risk, since mobile homes have lower damage than permanent homes and yet residents of mobile homes are at much greater risk. And hazards researchers know that property damage estimates are notoriously inaccurate, limiting the precision of any inferences we wish to draw. Indeed, this approach, like the others we use to see if the 2011 fatality totals resemble recent patterns of lethality, is imperfect. The inability of any one approach to provide a perfect perspective on how deadly 2011's tornadoes should have been suggests the value of several different approaches, to see if a consensus emerges from the several approaches.

As we discussed in Simmons and Sutter (2011), the SPC archive reported damage until 1995 in order of magnitude intervals, while tornadoes since 1996 have had dollar-value estimates of damage. Thus the damage reports are not consistent with each other, and reports from 1995 and earlier offer precision only within an order of magnitude. Consequently, we will conduct our analysis here using only tornadoes since 1996. In addition, many tornadoes in the SPC archive have no reported damage. Some of these tornadoes likely did no appreciable damage to real property, while others represent events in which damage occurred but no damage was reported. Distinguishing no reported damage from zero damage is practically impossible from the records, and in our analysis we focus on tornadoes rated EF-3 or stronger, tornadoes that are likely to have caused some property damage. Treating missing damage reports as tornadoes with zero damage would skew the ratio between fatalities and damage. Consequently, we omit in this analysis any tornado with no reported damage.

We first consider tornadoes between 1996 and 2010 to establish a baseline for comparison to 2011 tornadoes. Table 3.5 reports totals for these tornadoes, for EF-scale categories 3 through 5. The number of state tornado segments included in the totals is reported, along with the number of tornadoes in each category excluded due to no damage reported. All EF-5 tornadoes over the period are included as well

TABLE 3.5. Damage and Fatalities, 1996–2010

EF-Scale	Tornadoes	Fatalities	Damage	Damage per Fatality	Damage per Tornado	Tornadoes with No Damage Reported
5	7	118	$2,148	$18.2	$306.8	0
4	73	168	$3,285	$19.6	$45.0	31
3	390	374	$5,469	$14.6	$14.0	23

Damage amounts are in millions of 2011 dollars.

as over 95% of EF-3 tornadoes, but about 30% of EF-4 tornadoes are excluded. Column three reports the number of fatalities, while the next two columns report total damage and damage per fatality, both in millions of 2011 dollars.[2] A lower damage-per-fatality figure represents a greater level of lethality for a given amount of damage. Examination reveals that all three categories result in about one fatality for every $15 to $20 million of property damage over these years. The lowest value is actually for EF-3 tornadoes, but the difference between the categories is not large relative to the limits of precision in the damage data, and so we will not stress the difference. The final column of Table 3.5 reports damage per tornado in each category. Regression analysis reveals that the EF-scale rating of a tornado is the most important determinant of expected fatalities, and yet EF-4 and EF-5 tornadoes are not more deadly per million dollars of damage than EF-3 tornadoes. We see, however, that a substantial difference in damage per tornado does exist, with EF-5 tornadoes averaging $300 million in damage, compared with $45 million and $14 million for EF-4 and EF-3 tornadoes, respectively. The greater lethality of EF-5 tornadoes appears in this sample to be due to the greater damage to property, not greater lethality per million dollars of damage to property.

We can now evaluate the 2011 tornadoes for which we have damage estimates in light of the past relationship between damage and fatalities. We have data on U.S. tornadoes through April and from the May 24 outbreak from the SPC. The 2011 data are available based on county

---

2. Inflation adjustments are made using the CPI-U price index, with conversions to the August 2011 value.

TABLE 3.6. Damage and Fatalities, 2011

EF-Scale	Segments	Fatalities	Damage	Damage/ Fatality	Projected Fatalities	Segments with No Damage
5	9	237	$3,253	$13.7	178.7	2
4	19	131	$3,865	$29.5	197.6	10
3	42	50	$367	$7.3	25.1	7

Damage amounts are in millions of 2011 dollars.

tornado segments as opposed to state tornado segments, and thus we have damage, fatalities, and EF-scale rating for each county struck by a tornado as opposed to aggregated to one entry for all counties in the tornado path in a state. This difference in aggregating tornado track data should not render the comparisons we wish to make invalid. Again, some tornado segments have no reported damage and we omit these from our tabulations. Table 3.6 reports the totals for 2011 by EF-scale category. The first two columns report the number of included and excluded county segments, and the majority of segments in each category have damage reports, although about one-third of EF-4 segments are excluded. The next two columns present the fatality and damage totals for 2011 for the included segments, and the magnitude of the 2011 season is readily apparent in comparison with Table 3.5. Damage in 2011 in EF-4 and EF-5 tornadoes exceeds the inflation-adjusted totals from the prior 15 years, with EF-5 damage about 50% greater than the past 15 years. Damage in EF-3 tornadoes is substantially lower than over the past 15 years, which is not surprising as the included total of county segments is only about 10% of the 15-year total.

The final two columns of Table 3.6 present damage per fatality for 2011 and projected fatalities based on damage per fatality from Table 3.5, and these figures provide perspective on 2011 fatalities. Damage per fatality diverges from the 1996–2010 totals in each category, but the directions of deviation differ. EF-5 tornadoes resulted in one death per every $14 million in damage in 2011, which is slightly lower (indicating greater lethality) than the recent average of $18 million, but this difference is not large given the limitations of the damage data. Damage per fatality was $30 million in EF-4 tornadoes in 2011, indicating relatively low lethality, while one fatality occurred for every $7 million damage for EF-3 tornadoes, which is about half of the 1996–2010 average. Projected fatalities based on historical averages provide another way to compare lethality. Predicted fatalities are lower than actual fatalities for EF-5 and EF-3 tornadoes, but greater than observed for EF-4 tornadoes. Overall, though, 402 fatalities would be projected for the included tornado segments based on historical fatality rates, which differs by less than 5% from the actual total of 418. Again, given the limitations of the damage data, this difference is insubstantial.

Injuries provide a final perspective on fatality totals in 2011. Casualties provide a measure of the number of persons caught in the damage path of a tornado. While we would certainly expect damage and the number of persons injured or killed to be correlated, differences can exist due to buildings that were not occupied at the time of the tornado or variation in the average value of buildings damaged. And the imperfection of the damage data suggests the utility of alternative measures of impacts. Injury data, however, are far less reliable than fatality totals, due to the confidentiality of medical records and possible differences in effectiveness in tracking down injury information. Consequently, injury analysis is itself limited as well. In particular, the tornado records for 2011 reported no injuries for several killer tornadoes in 2011. While most tornadoes, in fact, result in no reported injuries (8.9% of tornadoes between 1950 and 2010), tornadoes with more fatalities than injuries are very rare. Indeed, the greatest difference of fatalities in excess of injuries (since 1900) is 15 (27 deaths and 12 injuries) in the 1997 Jarrell, Texas, EF-5 tornado. The absence of reported injuries in several 2011 county segments (as many as 23 or 27 fatalities without an injury) must be the result of a lack of reported injuries as of this time. Consequently, we omit all 2011 county segments with fatalities but no injuries to avoid biasing the analysis.

Table 3.7 reports the totals for 2011 tornadoes. We perform tabulations by EF-scale category, and use the period 1986–2010 (the tornadoes used to estimate the main fatalities regression model used in Section 3.2) to calculate a baseline ratio of injuries to fatalities. The first two columns of the table report the number of included and excluded county segments. The proportion of excluded segments is relatively

TABLE 3.7. Injuries and Fatalities

EF-Scale	Injuries/ Fatality 1986– 2010	2011 Segments	Fatalities	Injuries	Injuries/ Fatality 2011	Projected Fatalities	Killer Segments with No Reported Injuries
5	9.79	5	203	1,405	6.92	143.5	3
4	12.61	17	135	2,169	16.07	171.9	6
3	12.59	23	43	530	12.33	42.1	7

high (about one-third of EF-4 and EF-5 segments), and so we cannot evaluate as many tornadoes as we would like to. The next column reports the injury-to-fatality ratio for tornadoes between 1986 and 2010. Historically, the casualty ratio has been between 10 and 13 injuries per fatality in each EF-scale category, with EF-5 tornadoes being slightly more lethal based on the lower ratio, although given the limitations of the data, the difference may not be consequential. The fourth and fifth columns of Table 3.7 report the fatality and injury totals in the included county segments, with the sixth column reporting the casualty ratio in 2011 tornadoes, and the final column projected fatalities based on the historical ratio and injuries observed in 2011. Injuries per fatality were very close to the historical ratio for EF-3 tornadoes, with the projected fatality total differing only by one from the observed total of 42. The injury ratio was lower in EF-5 tornadoes in 2011 than over the prior 25 years, but higher in EF-4 tornadoes. As a consequence, fatalities in 2011 EF-5 tornadoes exceeded the predicted total by about 40%, while the 2011 EF-4 total is about 20% lower than predicted. Overall, the predicted fatality total for these most powerful tornadoes of 358 is about 6% less than the actual total of 381.

3.4. Conclusion

The 2011 tornado season featured on April 27 the first tornadoes in the United States to result in over 50 deaths in 40 years. Less than a month later the United States experienced the first 100-fatality tornado in almost 60 years. These tornadoes have completely reshaped our expectations and perceptions regarding tornado casualties. We have used tornadoes from 1986 to 2010 to extensively analyze the determinants of tornado fatalities in our previous research. The deadliest tornado over these years killed 36, and about 30,000 tornadoes occurred over these 25 years, so the data set used in our analysis is a very large sample. The fatality totals seem so surreal relative to recent totals that they suggest some new factor must be at work here. This emphasizes the need to evaluate whether the lethality potential of 2011 tornadoes could have been anticipated based on recent U.S. tornado history.

Perhaps it comes as a surprise then that several types of analysis—using patterns of tornado fatalities, injuries, damage, and numbers of buildings damaged—can come close to projecting the death tolls actually observed in 2011. Out-of-sample projections using a regression model of fatalities suggest that a 100-fatality tornado could have been expected given the tornadoes of 2011. We are less concerned that the fatalities regression model projected that it was the northern Alabama as opposed to the Joplin EF-5 tornado that would have produced the death toll, given the inherent limitation of county-level analysis. Projections based on past damage and injuries indicated that the Joplin tornado could have produced a death toll of between 98 and 165, again roughly in line (especially given the imprecision and limitations of the data) with the actual toll of 159. No one approach to projecting fatalities is perfect, which is why we have sought multiple perspectives, but all suggest that 2011 could have been projected to be the deadliest season the nation had experienced in decades if we had known about the tornadoes that would come.

We began this chapter with the question of whether extreme weather—the number, strength, and paths of the 2011 tornadoes—could explain the historic death toll. Our answer is seemingly yes, meaning that given the factors that explain tornado fatalities in the United States in recent decades, the tornadoes of 2011 could have been expected to kill hundreds. An implication is that researchers do not necessarily have to search for new and unrecognized forms of societal vulnerability to explain 2011. That extreme weather appears to be responsible does not, however, imply that social vulnerability did not play a role, because such vulnerabilities are reflected in recent patterns of fatalities. Societal vulnerability undoubtedly played a role in the human impacts, and in the future these vulnerabilities could be reduced. We will return to the path forward for tornado impacts research in the concluding chapter.

4

DOPPLER RADAR, WARNINGS, AND ELECTRIC POWER

The National Weather Service (NWS) underwent a thorough modernization in the early 1990s involving many changes and improvements (see Friday 1994 for details). The components of the modernization included an increase in the proportion of meteorologists employed, a consolidation and upgrading of the number of Weather Forecast Offices (WFOs) across the nation, and the installation of new computer software with improved graphics for forecasters. The centerpiece of the modernization of the NWS was the deployment of over 100 new Doppler weather radars (Weather Surveillance Radar, or WSR-88D) and the linking of the radars together into the first nationwide weather radar network. The new Doppler radars immediately improved the skill of tornado warnings (Polger et al. 1994; Bieringer and Ray 1994).

Our research has attempted to measure the effect of Doppler radar and tornado warnings on casualties. This analysis comprised the core of *Economic and Societal Impacts of Tornadoes*. The data set we have used in this research includes tornadoes since 1986, and as discussed in Chapter 3, the deadliest tornado in this data set resulted in 36 deaths. Three tornadoes in 2011 surpassed this total, with death tolls of 64, 72, and 159. The question naturally arises, then, if the addition of 2011's tornadoes to our analysis overturns our results about the effectiveness of Doppler radar or tornado warnings. The year is not over as we write this, and the Storm Prediction Center (SPC) obviously has not added 2011 tornadoes to its archive yet. In this chapter we use some preliminary data to extend our Doppler radar analysis through the deadliest of the 2011 tornadoes, even though the data to update results on tornado warnings are not yet available. We discuss the effect of extending casualty analysis on a preliminary basis into 2011 and offer some discussion of the challenges with evaluating how warnings and power outages across northern Alabama and Mississippi on April 27 might have affected fatalities.

4.1. Do Doppler Radar Effects Need To Be Revised?

To attempt to answer this question, we must extend the data set we used previously (with tornadoes through 2004) through 2011. We updated our data set using SPC archived tornadoes for 2005 through 2010, giving us tornadoes over the years 1986–2010, which were used in Chapter 3 to estimate fatality models for out-of-sample projections for 2011's major tornadoes. Validated tornado reports are still being assembled from NWS offices across the nation for the 2011 tornadoes, and so inclusion of 2011 tornadoes to the data set is preliminary. Greg Carbin kindly provided us with validated tornado reports being assembled for the SPC archive for all tornadoes nationally through April and from the May 24 tornado outbreak. We then constructed an entry for the Joplin tornado based on the NWS storm survey. The 2011 records are for county tornado segments, meaning that they have not yet been assembled into multicounty state tornado segments consistent with previous years of the SPC archive. While this is seemingly a minor point, it does reduce the number of fatalities for the long-track tornadoes of April 27. The largest county total of fatalities through April is 44 for Tuscaloosa, and this is the only county segment exceeding 27, although we do have the 159-fatality total for Joplin. Our 2011 records include 542 of the year's 552 fatalities to date.

Commercial damage in Tuscaloosa, Alabama, from the April 27, 2011 tornado outbreak. Nearly 11,000 buildings were damaged in the EF-4 Tuscaloosa–Birmingham tornado.

Residential damage in Tuscaloosa, Alabama, from the April 27, 2011 tornado outbreak.

Residential tornado damage to a neighborhood located northwest of Birmingham, Alabama, after the April 27, 2011 outbreak.

Search and rescue efforts in Tuscaloosa, Alabama, after the April 27, 2011 tornado outbreak.

Vehicular damage from the EF-5 tornado that struck Joplin, Missouri, on May 22, 2011.

Damage to Joplin High School, Joplin, Missouri, which was destroyed by the May 22, 2011 tornado.

St. John's Mercy Hospital in Joplin, Missouri. Five patients and one visitor were killed when the tornado hit the building. The remaining structure is being demolished, while a new hospital is constructed 3 miles away.

Community recovery efforts near St. John's Mercy Hospital after the Joplin, Missouri, tornado.

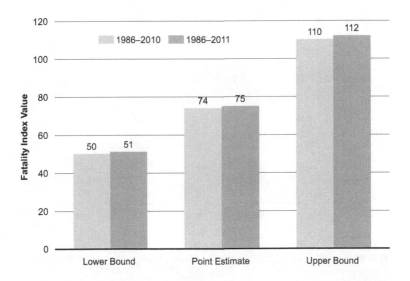

FIGURE 4.1. The effect of including 2011 tornadoes on the effectiveness of Doppler radar

We essentially want to see how adding the 2011 tornadoes affects the estimates of the determinants of fatalities. As discussed in Chapter 3, several tornado records from 2011 were missing injury data, and so we do not report an injuries regression that includes 2011 tornadoes. Our data set through 2010 includes over 28,000 state tornadoes with over 1,400 fatalities; as such, we already have a large data set in place, which would lead one to not expect a major change in the regression results. But the addition to fatalities is substantial, and thus inferences regarding the determinants of fatalities may change.

The full regression specifications are available in the supplemental online material for this book. Figure 4.1 displays the results for our variable of primary interest, Doppler radar. We display the point estimate of the effect of Doppler along with the upper and lower bounds of the 95% confidence interval, from regressions including and excluding 2011 tornadoes. The figure reports an index of the Doppler radar effect relative to tornadoes prior to the installation of the Next Generation Weather Radar (NEXRAD) network. The point estimate through 2011 indicates a 25% reduction in fatalities for tornadoes after installation

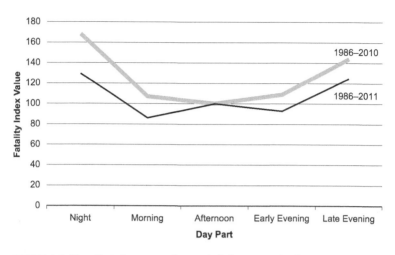

FIGURE 4.2. The effect of 2011 tornadoes on lethality across the day

of Doppler radar, which is a smaller effect than the 34% reduction through 2004 reported in Simmons and Sutter (2011, p. 140). But the point estimate when the model is estimated with tornadoes through 2010 is a 26% reduction in fatalities attributable to Doppler radar, so 2011 tornadoes result in a very minor change in the inference regarding the effect of Doppler radar. The change in the lower and upper bounds of the confidence intervals also indicates the very modest effect from inclusion of 2011 tornadoes.[1]

Although the addition of 2011 tornadoes has little effect on the estimated impact of Doppler radar, some differences do emerge. Perhaps the most noteworthy is regarding the effect of time of day on fatalities. Figure 4.2 reports an index of lethality based on tornadoes at the point estimates for different times of the day. The index is constructed

---

1. The difference in the estimated effect of Doppler radar here compared with our results in *Economic and Societal Impacts of Tornadoes* is due in small part to inferring path characteristic variables via linear interpolation from the 2000 Census and American Community Survey and the addition of extra years of data. The largest change in the estimate of Doppler occurs with the addition of 2005 tornadoes.

so that afternoon tornadoes have a value of 100 for each set of regressions. The estimates based on tornadoes over 1986–2010 substantially resemble the patterns discussed in Chapter 2 and *Economic and Societal Impacts of Tornadoes*: Tornadoes during the overnight hours are substantially more deadly than daytime tornadoes, with late-evening tornadoes reflecting an intermediate level of lethality. Here the point estimates indicate that overnight and late-evening tornadoes (12–6 a.m. and 8–12 p.m., respectively) are 68% and 44% more deadly than afternoon (12–4 p.m.) tornadoes, and these differences are statistically significant. Morning and early evening tornadoes are less than 10% more lethal than afternoon tornadoes, but this difference is not statistically significant. When 2011 tornadoes are added, afternoon tornadoes became relatively deadlier. Night and late-evening tornadoes are still more dangerous, but the point estimates indicate only 29% and 25% deadlier, respectively, now than afternoon tornadoes. Morning and late-afternoon tornadoes are now slightly (but insignificantly) less deadly than afternoon twisters. Late-evening tornadoes are now about as deadly as overnight tornadoes. The two very deadly Alabama tornadoes occurred during the afternoon hours, and this has altered the time-of-day inferences. Statistically significant differences are observed for several economic and demographic variables as well. Increased income now significantly increases fatalities, whereas it was insignificant through 2010, while the nonwhite population of a storm path has a greater marginal effect on casualties (a larger minority population now decreases expected fatalities more than through 2010). The elderly population of a storm path no longer significantly reduces fatalities, while a higher poverty rate now increases expected fatalities, whereas through 2010 this variable was insignificant.

4.2. Warnings and Power Outages

The 2011 tornado season raises at least two other interesting challenges for casualties analysis: whether the effectiveness of NWS tornado warnings has declined and whether power outages from earlier tornadoes and severe thunderstorms contributed to fatalities on April 27.

We discuss these issues here, although the data needed to satisfactorily address these issues are not currently available.

We have previously documented that tornado warnings save lives. Many of the killer tornadoes of 2011 were warned for; for example, the Joplin tornado was warned with a 17-minute lead time, and the entire path of the storm through Joplin was contained within the Storm Based Warning (SBW) polygons (NWS Service Assessment 2011). If well-warned-for tornadoes resulted in so many more fatalities than other recent, warned-for violent tornadoes, can we still claim that warnings save lives? Several responses are possible to this observation. Hazards researchers do not get to conduct controlled, randomized trials, so we will never see how many lives would have been lost if, say, the Joplin tornado had occurred without warning. We attempt to surmount this challenge by assembling a large data set of tornadoes, some of which happened to be warned for and others of which were unwarned. But many times the most powerful tornadoes are warned for, in part because violent tornadoes typically occur on days with numerous tornadoes and are spawned by supercell thunderstorms that do not escape detection by Doppler radar and NWS forecasters. The success of the NWS in warning for the strongest tornadoes can create analytical problems because the EF-scale rating is a relatively coarse measure and because tornado-path variables are based on county averages. Thus there may be variation in the potential lethality of tornadoes that our control variables do not necessarily capture well and that correlates with the number of observed fatalities. If so, regression analysis might underestimate the effect of warnings on fatalities. Because 2011 tornadoes were so deadly and well warned for, regression analysis updated to include these tornadoes might underestimate the effect of warnings.

We can offer an observation on warnings in 2011 concerning fatalities in homes in Joplin versus earlier EF-5 tornadoes. All of the earlier EF-5 tornadoes reported in the buildings-damaged-level analysis in Table 3.3 were warned for, and of the 110 fatalities in buildings in these tornadoes, 108 occurred in homes (permanent or mobile). More fatalities in Joplin occurred in other buildings than in permanent homes, and no fatalities occurred in mobile homes. The location of 11 Joplin fatalities is still unknown as of this writing, and even if we assume that these deaths all occurred in permanent homes, this would be a total of 76 deaths in homes. Damage reports do not always allow us to distinguish between homes and other buildings, and so we will just use the total number of buildings damaged or destroyed as reported in Table 3.4 for this comparison. With these assumptions, home fatalities-per-building damaged or destroyed are actually lower in Joplin (0.010) than for the other recent EF-5 tornadoes (0.013). If we evaluate warning response based on fatalities in homes, the response in Joplin does not appear to be worse than for previous EF-5 tornadoes. The issue of the larger number of business and hospital and nursing-home fatalities remains for Joplin, but residents in their homes appear to have responded in a similar manner to earlier tornadoes.

One reason offered for the death toll in the April 27 super tornado outbreak was power outages caused by severe thunderstorms and tornadoes that cut across Alabama, Mississippi, and Tennessee during the overnight and morning hours of April 26–27 (Samenow 2011). The overnight and morning outbreak on the 27th was a sizable event itself, with more than 70 tornadoes reported having produced 4 fatalities and 88 injuries. Over 250,000 customers were without power in Alabama alone after the morning storms. Without electric power, many households would have no access to television or the Internet and would be unable to charge cell phones. The disruption of standard communications could easily interfere with the dissemination of watches and warnings for the afternoon tornadoes. Warnings save lives, but the warnings issued by the NWS are just one component of the warning process, and with this process disrupted by a power outage, the life-saving benefits of warnings could be lost (AL.com article).

The impact of electric power outages on casualties is readily testable using our regression analysis. The difficulty exists in identifying whether electric service was indeed disrupted for a sufficiently large set of tornadoes for meaningful regression analysis. The data set we have used to examine the impact of warnings or Doppler radar on casualties contains over 20,000 tornadoes. We need literally thousands of tornadoes to attempt to tease out the effect of different variables on casualties. Thus we would need to establish the electricity (and perhaps

TORNADO SHELTERS—STILL NOT COST EFFECTIVE

Our first research on tornadoes was an attempt to quantify the benefits offered by tornado shelters and safe rooms. In *Economic and Societal Impacts of Tornadoes* we assessed the benefits and costs of shelters in detail and found that shelters offered cost-effective protection for residents of mobile homes in the most tornado-prone states, while shelters were not cost effective for residents of permanent or single-family homes, even in the most tornado-prone states. Do we need to revise these assessments based on the 2011 tornado season?

Actually no. To see this, let's consider the case of Alabama, which suffered 242 fatalities in 2011, including 234 on April 27. Between 1950 and 2010, Alabama had 369 tornado fatalities, or 6.05 per year. With the 2011 fatality total added, Alabama's total rises to 611 over 62 years, or 9.84 per year. Over the period 1985–2010, 32.0% of tornado fatalities occurred in permanent homes, and these are the fatalities that could be prevented with tornado shelters in single-family homes. If we assume that shelters in all single-family homes in Alabama *would* indeed prevent all of these fatalities, then shelters would prevent 3.15 fatalities per year. This is the benefit side of the calculation.

On the cost side, every permanent home in Alabama would need to be equipped with a shelter. According to Census Bureau estimates, Alabama has 1.45 million housing units in the "1, detached" category, which basically are single-family homes. Underground shelters large enough to protect the typical family cost $2,500 or more, while above-ground safe rooms cost in excess of $5,000. We use $2,500 as the cost of a shelter, recognizing that many households would likely pay more. The cost of equipping every single-family home in Alabama with a shelter is $3.6 billion.

The shelters, once built, will provide protection for many years to come; we will assume 50 years in these calculations. And since benefits and costs are not realized at the same time—the costs are incurred now while lives are saved for 50 years into the future—we must apply an interest rate to calculate the present value. We use a 3% real interest rate, which although low as of right now is a reasonable estimate of the long-run, risk-free interest rate.

Putting all of this together, we arrive at a cost-per-life saved of $43.4 million. We have used many assumptions in arriving at this figure; for a further discussion of the method applied and how the assumptions affect the cost-per-life saved, see Chapter 5 of *Economic and Societal Impacts of Tornadoes*. This cost-per-life saved can be compared with estimates of the value of a statistical life generated by economists over the years. Estimates from many different market settings indicate that this value is less than $10 million. A person with a statistical value of life of $10 million would find safety measures that save lives for less than $10 million to be good investments, while measures with a cost above this amount would not be attractive. By this metric, tornado shelters do not offer cost-effective protection for single-family homes, even in Alabama.

Our analysis here based on the cost-per-life saved should not be interpreted prescriptively—we do not mean to imply that people *should* or *should not* invest in a storm shelter or safe room. Protecting oneself and one's family is a personal decision based on the value we each place on safety, our fear of tornadoes, and the value of the peace of mind offered by a shelter or safe room. This calculation illustrates for researchers and policymakers that we should not expect widespread purchase of shelters. Tornado safe rooms do not appear to be a cost-effective way to reduce fatalities, even in the aftermath of the 2011 season.

other utilities like cable or satellite television) service status not just for the tornadoes on April 27 but for thousands of other tornadoes. This task is beyond the scope of this research project. Establishing the blackout status for the individual tornadoes in the April 27 outbreak would not allow a good econometric investigation of this question because there would be too few data points to estimate a robust regression model, making the inferences questionable. Power outages are not unique to the April 27 outbreak, as the thunderstorms accompanying a large tornado outbreak will often knock out power in advance of tornadoes. Consequently, the patterns of casualties in regression analysis will incorporate in total the effect of power outages.

To investigate whether the earlier tornado outbreak on the 27th might have affected casualties in the afternoon and evening outbreak, we identified county tornado segments in Alabama, Mississippi, and Tennessee during the later outbreak. We then partitioned this group of tornadoes based on whether these counties had a reported tornado during the late evening of the 26th or before noon on the 27th, and then tabulated fatalities and injuries per tornado by EF-scale rating. Table 4.1 reports the totals for each category and in total. About half of the strongest county segments (EF-3 and stronger) occurred in counties that had morning tornadoes (27 of 58), and so we have good variation of prior tornado status that could allow us to detect a difference in casualty rates. The injury totals are presented but should not be considered very reliable due to the county segments with missing injury reports discussed in Chapter 3. Table 4.1 provides evidence that morning tornadoes might have increased the lethality of the afternoon outbreak. Fatalities per county segment are higher in the counties that experienced morning tornadoes for the EF-4 and EF-5 categories, but not for EF-3 tornadoes. Overall, 3.2 fatalities occurred per county segment where tornadoes had occurred in the morning compared with 1.1 fatalities per segment in counties that had not experienced an earlier tornado. The earlier tornadoes could have affected lethality in a variety of ways, including power outages, by diverting residents' attention from the later severe weather threat or through perceptions that the tornadoes in the morning would prevent tornadoes later in the day.

TABLE 4.1. Morning Tornadoes and Afternoon Casualties, April 27, 2011

EF-Scale	Morning Tornado in County			No Morning Tornado in County		
	Segments	Fatalities/ Segment	Injuries/ Segment	Segments	Fatalities/ Segment	Injuries/ Segment
5	3	13.00	12.33	6	10.33	17.67
4	15	8.67	130.9	8	2.13	7.63
3	9	0.89	3.89	17	1.71	15.18
2	5	0.60	0.00	16	0.13	0.88
1	20	0.05	0.20	24	0.04	0.17
0	4	0.00	0.00	26	0.00	0.04
Total	56	3.23	36.41	97	1.14	4.58

Observations are county tornado segments after 12:00 p.m. local time on April 27 in Alabama, Mississippi, and Tennessee. A morning tornado is a tornado that struck the county before 12:00 p.m. on April 27 or in the late evening on April 26.

The regression model we estimated with 2011 tornadoes in Section 4.1 allows us to provide additional evidence on the role of the earlier morning tornadoes on the lethality of the April 27 outbreak. The predicted values from the regression allow us to evaluate how deadly a tornado should have been given the factors we can control for with the available independent variables. The residuals from the regression—the differences between predicted and actual fatalities—can provide an indication of other factors that might affect lethality. Any important factor not controlled for through the included variables can result in a large residual, or the regression model significantly over- or underpredicting fatalities. We use predicted fatalities for the afternoon tornadoes in Alabama, Mississippi, and Tennessee, the same tornadoes considered in Table 4.1, and compare in Figure 4.3 predicted with actual fatalities for tornado segments in counties that did and did not have morning tornadoes. The results reinforce the differences observed in Table 4.1. For the 56 segments in counties with morning tornadoes, 133 fatalities were predicted but 181 observed, while 142 fatalities were predicted and only 113 observed in counties where tornadoes did not occur in the morning. "Second wave" tornadoes were deadlier than

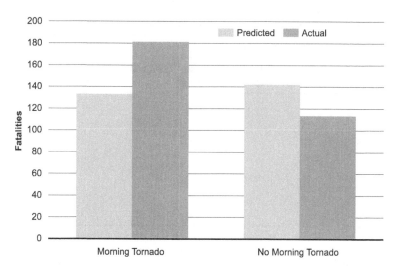

FIGURE 4.3. Early tornadoes and fatalities on April 27

would have been predicted given tornado and path characteristics, which suggests that the disruption due to morning tornadoes may have contributed to afternoon fatalities.

4.3. Conclusion

Does the historic 2011 tornado season affect inferences about the effect of NWS efforts to reduce tornado casualties? Addition of 2011 tornadoes to our regression analysis appears to significantly alter the inferences regarding a number of determinants of fatalities. It turns out that the inferences regarding Doppler radar on fatalities is not really affected by extending our data set to include 2011 tornadoes, but the time-of-day effects are substantially altered. Tornadoes after dark are still more deadly, but their relative lethality is reduced, and afternoon tornadoes are relatively more deadly than in previous regression analyses. We do caution readers that this regression analysis is preliminary and based on tornadoes only through April plus selected May tornadoes.

The data are not yet available to reevaluate the effect of 2011 tornadoes on warnings. The NWS performed well in the 2011 tornado outbreaks, and so the killer tornadoes were likely well warned. If all of the worst killer tornadoes are warned for, it can be difficult to identify the effect of warnings or lead time on fatalities, because statistically all the deadliest tornadoes were warned for and we do not (fortunately) observe the number of lives lost if these most dangerous tornadoes occurred without warning. The broad similarity of historical ratios of damage or injuries to fatalities to those in the 2011 tornadoes suggest that warning response probably was not markedly worse in 2011 and that warnings will probably not be substantially less effective after 2011 is added to the analysis. The record-setting outbreak across Alabama, Mississippi, and Tennessee on April 27 occurred after severe thunderstorms and tornadoes moved across the area during the early morning hours, knocking out power to thousands of customers. We have found that afternoon tornadoes in counties struck by tornadoes in the morning are more deadly, both in univariate comparisons and using predicted fatalities from a regression model. This evidence is far from definitive, but it suggests that multiple rounds of tornadoes in the same area could lead to an elevated risk to life and limb if power outages are caused by earlier thunderstorms and tornadoes.

RECOVERY FROM TORNADOES

5

As tornado researchers who had just published a book on tornado impacts, we received a number of calls from reporters during the tornado season. As often as not, the topic they wanted to know about was whether communities recovered from disasters like tornadoes. Without a prompt recovery, the disaster lingers and continues to inflict economic pain even as the emotional scars begin to heal.

Natural hazards researchers have debated among themselves about what even constitutes a successful recovery. One might think that the definition is obvious, a restoration of the status quo ante, or as close as physically possible and desired by residents. This would include the removal of debris, repairs to damaged buildings and infrastructure, and demolition and reconstruction of buildings beyond repair. But some scholars in the natural hazards literature argue that recovery should go beyond simply rebuilding and make communities more resilient to disasters (Burby 1998; Mileti 1999). One reason for this is that a community might have excessive exposure to hazards, and if so, rebuilding the status quo will simply perpetuate the vulnerability. A community may not have recovered from a flood if it is still as vulnerable to a flood as before, and repairs and rebuilding will be necessary again in a few years. The aftermath of a disaster may represent the best, and perhaps only, time to redevelop a community to reduce vulnerability. Buildings and infrastructure once built and in place cannot be moved, making relocation very costly. If an entire neighborhood or business district has to be rebuilt after a disaster, the costs of relocating will be temporarily low. Quickly restoring the status quo ante merely perpetuates the same vulnerability that produced the current disaster. If so, policymakers may even want to take steps to slow or even temporarily halt recovery.

We will stick with the narrow view of recovery here and consider recovery to be successful if the community returns and remains at its state prior to the tornado. This seems like a reasonable approach for tornadoes because to our knowledge areas of high local vulnerability akin to flood plains do not exist. We begin by considering some different ways a tornado can impact a community and permanently alter its economic health and reviewing some prior research on tornado impacts and recovery. We then present some new evidence on long-run effects of tornadoes on communities, using both population and retail sales measures of health and performance. The good news for the communities struck by tornadoes in 2011 is that most communities have shown few long-run effects from major tornadoes in the past.

5.1. Disaster Impacts and Evidence on Recovery from Tornadoes

Natural disasters can have a community-wide impact, creating a policy-relevant dimension to protective actions (Burby 1998). While a fire is a traumatic event for a family or business, it has little impact on the larger community. Community-wide impacts arise due to the spillovers that result from losses and the concentration in time of various losses. A temporary closure of one store or restaurant or school will be an inconvenience, but destruction of a town's entire business district will significantly affect residents whose homes were not in the tornado's path. When a disaster levels an entire neighborhood, residents face uncertainty about whether neighbors will rebuild and businesses reopen. If others choose not to rebuild, a resident could be rebuilding into a substantially changed neighborhood. Reconstruction becomes a coordination game in which residents and businesses will want to rebuild only if others choose to rebuild as well.

A natural disaster can adversely affect the economic health of a community in three ways. The first is through the potential relocation of households and businesses directly affected by the disaster. Destruction of a residence or business reduces the cost of moving as a new structure must be built. Households or firms that wanted to relocate prior to the tornado but whose gain from moving did not exceed the cost, or who would have benefitted from relocating but had not done so due to procrastination, will relocate after the disaster. Businesses may decide not to reopen after a disaster, either because losses from the tornado render operation unprofitable or the owners may have planned to close the business and retire soon. Families who lost a loved one may decide not to rebuild at their previous location.

The second effect is through community-level or second-round impacts. A household that was either not directly affected by the disaster or would plan to rebuild in place may decide to move because of the departure of friends and neighbors. Businesses may be able to rebuild but unable to recover if the disaster reduces its local customer base. The departure of businesses may reduce employment or shopping opportunities and lead households to move. The initial decisions of businesses or households directly affected by the disaster cause impacts for others not directly affected or those able to recover initially. Community-level or spillover effects can result then in a second (or third) wave of impacts. The key factor for many spillover effects is not whether the residents or businesses in place before the disaster return but whether the buildings get rebuilt and occupied.

A third effect arises through updating of hazards risk, with the affected area now being perceived as more dangerous and consequently a less desirable place to live or work. Galveston, Texas, was the second-busiest port in the Gulf of Mexico before the 1900 hurricane devastated the city, and this vulnerability led to a relocating of port facilities in Houston (Larson 1999). This effect can lead residents outside of the tornado path to move or deter others from moving to the area, creating a community-wide effect. Indeed, by providing new information about natural hazards risk, a disaster can have very broad effects across a state or region, wherever people decide risk is now too great. A very broad impact due to a region-wide perception of heightened risk can be difficult to detect, as plausible control communities outside of the disaster area can be affected by risk-perception effects.

Government actions can also affect recovery after a natural disaster. The public sector typically is responsible for debris removal and rebuilding of infrastructure, and speedy accomplishment of these tasks can facilitate recovery. The reopening of public schools can be an important component as well, both as a part of life returning to normal and because families are more likely to return to their old neighborhood if they know that their children will be able to return to a community school with their classmates. Government, however, can also interfere with recovery. Hurricane Katrina illustrates that both the immediate response (Sobel and Leeson 2007) and long-term recovery (Chamlee-Wright 2010) can be impaired. New Orleans went through multiple rounds of planning for the rebuilding process, with different plans at different times proposing to restrict rebuilding in some neighborhoods. The rules of the game for rebuilding kept changing on residents, and the resulting uncertainty can prevent residents from overcoming the coordination problem inherent in rebuilding a neighborhood or community. Delay might be inevitable if government seeks to guide rebuilding in a way to reduce vulnerability to future disasters, but policymakers need to be cognizant of the potential disruption that can result from delayed rebuilding.

The regional economic effects of tornadoes and hurricanes have been examined by economists. We have already reviewed some of this literature in *Economic and Societal Impacts of Tornadoes* (see also Ewing et al. 2007) and we will only discuss the most relevant research here. In general, stronger hurricanes have a greater impact on employment than weaker storms (Belasen and Polachek 2009). Tornadoes are much smaller events than hurricanes, but powerful tornadoes typically occur as part of large outbreaks, which can result in damage across a region. Several studies have specifically investigated the economic impact of and recovery from tornadoes. Ewing, Kruse, and Wang (2007) examined housing price indices in metropolitan statistical areas affected by windstorms, including tornadoes in Nashville (1998), Oklahoma City (1999), and Fort Worth/Arlington (2000). Tornadoes caused a decline in the price index of between .4% and 1.8% with the declines ceasing by the fourth quarter after the event. Real estate values are significant for economic recovery, particularly for states that rely more on property taxes to fund local government and schools.

Studies of the labor market reaction to the 1999 Oklahoma and 2000 Fort Worth tornadoes provide insights on the effect of tornadoes on large local economies. In a series of papers Ewing, Kruse, and Thompson (2003, 2005, 2007) found that the 1999 tornadoes had a significant effect on only one sector of the central Oklahoma and southern Kansas labor markets. The results in Fort Worth were similar, with a reduction in growth in the labor market, but no contraction in overall employment and decreased labor-market volatility after the tornadoes. Thus, although tornadoes represent a serious threat to small economies, large metropolitan economies have usually been quite resilient to these storms in recent years.

5.2. Population Change after Significant Tornadoes

The past can provide a guide for recovery from the 2011 tornadoes. Have communities recovered from tornadoes in the past or have growth trajectories of the affected communities been altered? We provide some perspective on this question by examining the population effect of the worst U.S. tornadoes since 1900. We identify tornadoes that killed 20 or more persons from the Storm Prediction Center (SPC) archive for tornadoes since 1950 and from Grazulis (1993) for tornadoes prior to 1950. Population is the only measure of impact available over such a lengthy sample period and is a less than perfect impact measure. Population does reflect overall economic health, as a community with profitable, thriving businesses and attractive neighborhoods and schools will tend to attract and retain residents, while a community in decline will lose population. But moving costs are typically substantial, and

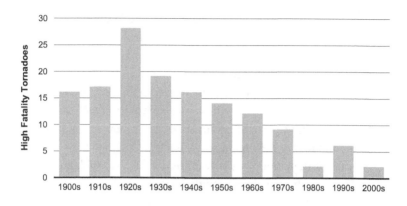

FIGURE 5.1. Tornadoes with 20 or more fatalities by decade

thus a tornado could significantly affect a community without triggering a population loss. And only county-level population is available for tornadoes so far in the past, and counties are large relative to tornado-damage paths. A significant very local impact could be small relative to the entire county, and a tornado could result in a substantial population relocation within the county, which we would not detect in this analysis. Nonetheless, the interest in recovery from tornadoes suggests the value of examining the available evidence, which happens to be only county populations for this set of tornadoes.

We identified 141 tornadoes that killed 20 or more persons over the period, and the paths of these tornadoes included 367 counties. We performed calculations based on the population of the entire tornado path and then for individual counties. Figure 5.1 displays the number of tornadoes meeting the 20-fatality threshold by decade, and not surprisingly, given the declining lethality of tornadoes over the 20th century, a majority of tornadoes occurred in the early decades. The 1920s had the most 20-plus-fatality tornadoes, and 68% (43%) of the tornadoes occurred before 1950 (1930).[1]

---

1. Grazulis (1993) reports fatality totals for entire storm paths while the SPC reports totals for state tornado segments, which also tends to reduce the number of tornadoes hitting the 20-fatality threshold since 1950.

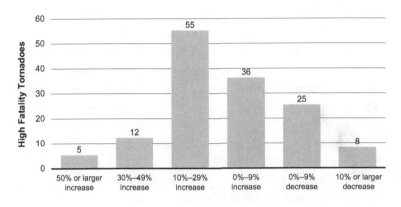

FIGURE 5.2. Distribution of tornado-path population changes after a killer tornado

The counties in the paths of these tornadoes typically experienced population growth during the decade in which the tornado struck. Figure 5.2 displays the distribution of population changes during the decade of the tornado. The average and median paths experienced 12% and 10% increases in population. Overall, 108 tornado paths experienced population increases and 33 experienced decreases, and 12% of tornado paths experienced an increase of 30% or more during the decade. Of course, the U.S. population has been growing since its founding, and so population increases might still be expected in counties struck by tornadoes. A more appropriate question might be whether the tornadoes affected the population growth rate of the affected communities, and to answer this question we also consider population change in the decade before the tornado struck. The tornado-path counties were growing prior to the tornadoes, with the mean and median paths experiencing 11% and 9% population increases, respectively.[2] A better measure of an effect on population growth or community health would consider the difference between the current and prior decade growth rates. Most tornadoes had a small effect on population growth, as the mean and median differences in population

2. We were unable to find county populations before 1900 and so tornadoes during the decade 1900–1909 were omitted from these tabulations.

change are 1 percentage point increases and decreases, respectively. Some very large reductions in population change rates occur in the sample, including an 80 percentage point decline for a tornado that struck Oklahoma County in 1930, from 90% growth during the 1920s to 10% growth in the 1930s. Typically little difference is observed, with 48% of the tornadoes resulting in a difference of 10 percentage points or less.

The tornadoes considered here did not impose equal impacts on their communities, as fatalities differed by more than an order of magnitude and county populations by several orders of magnitude. Impact could be measured in several ways, and here we consider the number of persons killed, fatalities per county in the tornado path (because a given number of fatalities represents a smaller impact if spread across five counties instead of one), and persons killed per 100,000 residents in the Census prior to the tornado. Table 5.1 displays the correlations of these measures of impact with population change, the difference in the population growth rate, and the 20-year population change (from the Census before the tornado to two Censuses after). All of the correlations are modest (less than .1 in absolute value) and all but one is negative, indicating that tornadoes with greater impact result in a greater slowdown of population growth. The largest correlations are for the number of fatalities, which suggests that the magnitude of the event (as opposed to impact relative to the damage path) is most closely associated with population change, although the differences between the correlations are also small. We might expect that effects on population change should be diluted in more populated communities, as a given level of impact will proportionally be more significant in a smaller county. The correlations in Table 5.1, however, remain essentially unchanged when calculated only for the tornado tracks with the smallest populations prior to the tornado.

We do observe a difference in population change when we compare the 35 tornadoes with the largest death tolls to the 35 with the smallest death tolls (35 is approximately a quartile of these killer tornadoes). Population increased 8.4% in the deadliest quartile of tornadoes during the decade of the tornado as opposed to a 17.9% increase among the quartile with the smallest fatality totals. These growth effects per-

TABLE 5.1. Tornado Impact and Population Change

Measure of Population Change	Measure of Impact		
	Persons Killed	Persons Killed per County In Tornado Path	Persons Killed per 100,000 Tornado-Path Residents
Population Change, Decade of Tornado	-0.088	-0.003	+0.018
Twenty-Year Population Change	-0.084	-0.010	-0.020
Population Change Decade of Tornado —Prior Decade Population Change	-0.095	-0.081	-0.060

Numbers are correlations of the variables.

sist over time, as the average 20-year population growth is 15.5% and 23.3% in these two quartiles, respectively. The growth rate declined 5 percentage points in the deadliest quartile of tornadoes compared with a 5 percentage point increase in growth in the least deadly quartile.

We repeated this analysis using the 367 counties in the paths of these tornadoes as opposed to aggregating to the tornado path. Again, we find no evidence of a significant impact on population growth. The mean and median county experienced 8% and 5% population increases, respectively; with these, change rates were reduced by about one percentage point from the previous decade. Fatalities have slightly stronger correlations with population change (as large as -0.14), but fatalities per county is essentially uncorrelated with population change. The differences in population change in path counties between the quartiles of deadliest and least deadly tornadoes are also very small.

5.3. Case Study: The Tri-State Tornado

The 1925 Tri-State Tornado is the deadliest tornado in U.S. history and resulted in 695 fatalities and over 2,000 injuries in a 219-mile track across Missouri, Illinois, and Indiana. The Tri-State Tornado is included in the set of tornadoes examined in Section 5.2. Considering population change only in the counties struck by the tornado, however, does not really inform us about the population change that might have been expected if the tornado did not strike. The counties in the tornado path might experience population decline, but for reasons that have nothing to do with the tornado. A declining regional economy, however, might result in population declines in the affected *and* surrounding counties. Or a region might be experiencing rapid growth and the tornado could reduce growth in the affected counties; yet the path counties could still grow after the tornado.

We use the Tri-State Tornado as a case study of the other factors in tornado-induced population change and widen the analysis to consider all 309 counties in the three states. Fourteen counties were in the path of the tornado. The population declined in 10 of these counties, while 4 counties experienced population growth (albeit an increase of one person in one county). The largest increase was 3,362 in Cape Girardeau County, Missouri, while the largest decrease was 7,212 in Williamson County, Illinois. The percentage changes ranged from an 11.3% increase in Cape Girardeau to an 18.4% decline in Hamilton County, Illinois. Overall, the population of the counties in the path decreased by over 16,000 in the 1920s, or 4.7%, with the average of the county changes being a 6.2% decline.

But what about population change in the other counties in these states during the 1920s? If other counties were similarly declining in population in the 1920s, it may well be inappropriate to attribute the decline to the Tri-State Tornado. Of the 295 other counties in the three affected states, 102 experienced population increases while 193 saw population declines during the 1920s. On net, the population of the rest of these states increased by 1.7 million persons (13.6%), but the mean and median of the population changes were -0.7% and -3.3%, respectively. The urban areas of the states grew during the 1920s, in cities such as Chicago, St. Louis, and Indianapolis, but more rural, agricultural counties lost population. The Tri-State Tornado path counties were largely rural, rendering the cause of the population decline in these counties unclear.

A regression analysis can provide some further insight on population change. To do so we estimate a simple cross-section ordinary least-squares regression with the percentage change in population during the 1920s as the dependent variable. Available Census data from 1920 allow the use of few other control variables. We are able to include the percentage of county population living in urban areas, the percentage of men in the county population, and the percentage of the population under age 10 in 1920 as controls. We include the percentage population change in the county over the prior decade, 1910 to 1920, to control for the local trend in population. And we include state dummy variables to control for state policies that might affect population change.

The regression results are available as part of our supplemental online material. We control in the regression for counties in the tornado path using a categorical or dummy variable. The regression indicates that the population of path counties was about 6.5 percentage points smaller in 1930 than other counties in the state (which is statistically significant). Among other significant determinants of population change, more urbanized counties grew faster, while population change from 1910 to 1920 persisted over time. Not all counties struck by the Tri-State Tornado experienced a similar level of impacts. Two counties—Jackson and Franklin, Illinois—suffered over 100 deaths in the tornado while four counties in the path had no deaths. A regression model that treats all path counties equally might obscure a relationship between tornado impact and population change. We therefore also estimate a regression in which only counties that suffered fatalities are counted as being affected by the tornado, and population growth in these counties was almost 9 percentage points lower in 1930, everything else equal. The population impact persists over time, as the path counties had almost 10 percentage points smaller population in 1940 than other counties in the state, although this result is only on the margin of statistical significance. Other extreme weather events during the decade, namely the Mississippi and Ohio Rivers flooding and other major tornadoes (a deadly tornado struck St. Louis in 1927) did not significantly alter population change in the affected counties, or the population change attributable to the Tri-State Tornado. Overall, it appears that the deadliest tornado in U.S. history did have a modest long-run effect on population growth in the affected counties.

DO TORNADOES CAUSE LOCAL ECONOMIC BOOMS?

After a disaster occurs and search-and-rescue efforts conclude, a community begins the rebuilding process—debris must be removed, homes and businesses and schools repaired and rebuilt, and temporary housing facilities found. As the pain of the event passes and recovery begins, people often observe an increase in economic activity in an area affected by a disaster: hotels and restaurants are very busy and retailers often experience increased sales. This surge in business activity often leads people to credit a tornado (or hurricane, earthquake, or other disaster) with precipitating a boom for the local economy. Does this mean that disasters can be good for the economy?

A tornado can increase local economic activity, but this does not mean that the local community is overall better off due to the destruction caused by the tornado. To properly assess the tornado-induced boom, we must consider three factors. First, disasters often divert spending, and we must keep in mind the spending that does not occur after a tornado. When events divert spending, we see the new spending that occurs but do not see and fail to recognize the spending that does not take place. Economist Frederic Bastiat labeled this the "broken window" fallacy. The fallacy arises because when the window of a store is broken, we see that this increases business for the company that replaces the window (and then has more money to spend), but we don't see the purchases that the store owner doesn't make because of the repairs. One of the tasks of economists is to remind people of the unseen or diverted economic activity. Families and businesses that have to replace clothing, furniture, and office supplies lost in a tornado will do so in part by diverting spending; for example, a family may forego a vacation this year due to the tornado. Also many of the resources that flow into a community after a disaster—insurance payments, state and federal disaster assistance, private charitable donations—reduce spending elsewhere in the economy.

Second, tornadoes alter the normal balance between stocks

(e.g., wealth) and flows (e.g., income) in the economy, rendering flow-based measures of economic well-being temporarily misleading. Many stocks involve durable goods like houses, office buildings, appliances, and medical equipment. The flow of economic activity required to maintain and grow a stock, like the housing stock, is small relative to the size of the stock; the number of new homes built in a year is small compared to the stock of homes. Gross domestic product (GDP) measures the flow of economic activity in a given period of time. Normally stocks and flows will be in balance, and then an increase in the flow of economic activity can be interpreted as a signal of a healthy economy. When a tornado destroys part of a community's housing or capital stock, a temporary increase in the flow of activity occurs to restore the stock. The increase in the flow of economic activity postdisaster is not consequently the same indication of economic prosperity as under normal circumstances. The increase in economic activity due to reconstruction is temporary and represents an inherent limitation of GDP as a measure of economic well-being.

Finally, disasters have distributional effects, as researchers have long recognized (Scanlon 1999) but do not always emphasize. A tornado makes many people worse off—those who lose homes or businesses and obviously those killed or injured—but can make others better off. The tornadoes of 2011 occurred after a long slump in the construction industry, and so struggling home builders will rightly regard the tornadoes as good for business. The diversion of spending contributes to these distributional effects; retail stores located outside of the damage path, for example, will experience increases in sales, which can result in overtime for employees and bonuses for managers. The flow of resources into a community (insurance payments, disaster assistance) also affects the distribution of gains and losses. And the destruction of homes and offices will affect the value of the remainder of these stocks. Thus many people in a community might honestly assess that they personally are better off as a consequence of a tornado, and this can create an impression that disasters are good for the local economy.

5.4. Tornadoes and the Local Economy

Population as mentioned is a relatively crude dependent variable to use to measure the impact of a tornado on a local economy. Households are likely to shoulder substantial impacts before moving, while the disaster may only trigger other moves that were likely to occur otherwise. Unfortunately, many measures of economic activity (like gross domestic product) are not calculated for small geographic areas, while others like unemployment are tabulated only for larger metropolitan areas. Unemployment has been used to explore tornado impacts in previous research, as discussed in Section 5.1. Here we turn to an alternative measure of local economic activity—sales of goods subject to state and local sales tax. Sales subject to sales tax provides a good measure of the health of a local retail sector, which is both an important employer and contributes to the quality of life. Although the sales lost after a tornado in one community are likely diverted to neighboring communities, the diversion of retail sales indicates a reduced level of convenience, reduced revenues for local government, and could portend future job losses for the local economy. A limitation to the uses of sales tax data is that not all states impose a sales tax, and if a state does not allow cities or counties to add on their own sales tax, sales subject to tax may not be tabulated for local communities. Nonetheless, sales subject to tax has been recommended as a good variable to use to analyze the health of local economies (Marshment and Rogers 2000).

We use a database of sales subject to tax for the counties in Oklahoma maintained by the Center for Economic and Management Research at the University of Oklahoma. We investigate the retail impact of the 16 tornadoes in Oklahoma between 1980 and 2010, which produced at least $40 million in reported inflation-adjusted property damage.[3] We calculate for each of the counties listed in the storm path the monthly average of seasonally adjusted sales subject to tax for the 6 months prior to the tornado and 12 months after, and then take the

---

3. Reported damage was less than $15 million in the remaining tornadoes in the state over the period, so these tornadoes are clearly the most damaging in Oklahoma over these years.

percentage difference of monthly sales after and before. This is a very simple comparison, but it allows us to ask if we observe any noticeable impact on a county retail sector after a tornado. The 16 tornadoes struck a total of 23 counties. Overall, tornadoes do not cripple a county economy, as mean and median changes in monthly sales are +5.0% and +5.2%, respectively. Sales subject to tax declined after the tornado in 5 of the 23 counties, with the decreases ranging from 1.3% to 12.9%. The greatest local economic impact was in Lincoln County after an EF-4 tornado on May 3, 1999, destroyed a popular outlet mall in Stroud, Oklahoma, which was never rebuilt.

Damage in these tornadoes ranges from just over $40 million to $1.3 billion (adjusted for inflation) in the May 3, 1999, EF-5 tornado that struck the Oklahoma City metropolitan area. Presumably the amount of damage affects the local economic impact, but the relationship is weak at best, as both damage and damage per county in the storm path are positively correlated (around +0.2) with the change in retail sales. The size of the local economy might also affect the impact of damage, as $1 million in property damage will result in greater disruption in a small economy than a large economy. Again, though, we find little evidence that damage or damage per county scaled by the sales subject to tax prior to the tornado influences damage, as the correlations with the change in sales are +0.10 and +0.19, respectively. The local retail sector may not show more of an impact because sales can simply be diverted to undamaged businesses in the community. While in some circumstances the economic impact can be significant, on average the local retail sector shows little adverse effect even from significant tornadoes.

5.5. Conclusion

After the initial rescue-and-response phase of a disaster, attention naturally turns to recovery and whether the affected communities will be able to rebound from the tragedy or whether the disaster might even be a blessing in disguise by stimulating the economy. Will the many communities ravaged by tornadoes in 2011 be able to recover and at least return to their prior level of economic activity and well-being? Natural hazards researchers have consistently found that the answer to this question is most always yes, despite the losses a community suffers and the potential for a coordination failure to slow or prevent recovery. Exceptions exist, and not every neighborhood or town will recover quickly, especially if the local economy was weak prior to the disaster. Existing research on recovery from tornadoes confirms this general point that the economic impacts of tornadoes are quite modest and recovery relatively rapid. We have looked at the population and retail sector impacts of major tornadoes and found them to be modest as well. This is perhaps not surprising because the damage path of a tornado is small relative to cities or metropolitan areas, which helps focus expectations of recovery. In part this is due to the indomitable human spirit and desire to improve our lives, which leads people to pick up the pieces and rebuild after disasters. This outcome was noted by economist John Stuart Mill (1848) over 150 years ago: ". . . What has so often excited wonder [is] the great rapidity with which countries recover from a state of devastation; the disappearance, in a short time, of all traces of the mischiefs done by earthquakes, floods, hurricanes, and the ravages of war."

6

LESSONS LEARNED AND THE PATH FORWARD

We began this book with the goal of determining the extent to which extreme weather and extreme vulnerability were responsible for the historic 2011 tornado death toll. Our analysis finds that extreme weather—meaning the number, strength, and location of the year's tornadoes—can explain much of the deadly season. We have employed a variety of approaches to reach this assessment, including out-of-sample projections based on regression analysis of tornado fatalities and applying recent ratios of fatalities to damage, buildings damaged, and injuries to the impacts of this year's tornadoes. While the different methods lead to slightly different projections, each approach implies that based on the impacts observed, we would have expected a death toll not seen in decades in the United States. Admittedly, each of the approaches we have used has its limitations, and this is why we have employed a variety of approaches.

Observers have offered a number of reasons or special vulnerabilities to explain why so many persons died in 2011, such as population growth in tornado-prone areas, poor warning response, power outages from storms the morning of the April 27 outbreak, and even climate change. We have not explicitly rejected any alternatives and certainly believe that the evidence suggests power outages did contribute to the April 27 death toll. The ability of recent patterns of lethality to project a season total of fatalities in excess of 300 suggests that special vulnerabilities are not needed to explain the season.

What does this analysis imply for future tornado seasons? Are we likely to see 100-fatality tornadoes and 500-fatality seasons regularly in the future? And do communities struck by this year's tornadoes have any hope of recovering? Since extreme weather appears to have driven the 2011 fatality totals, there is no reason to expect fatality totals will not revert to recent averages prior to 2011, which were around 60 deaths annually over the past 20 years. Also, evidence shows that communities appear to recover pretty quickly from tornadoes, as population and retail sales usually increase after a significant tornado. This probably should not come as a surprise, as tornado paths are small relative to counties or metropolitan areas, and thus recovery will not have large-scale coordination game elements seen for hurricanes, floods, or earthquakes. Retail shopping can be diverted and homes and businesses can often locate to vacant space in the community. Infrastructure need only be repaired, while streets do not have to be redesigned. As a result, there is a high level of expectation that devastated neighborhoods will return to normal. In addition, tornadoes do not reveal areas of exceptional vulnerability that need to be avoided in rebuilding, as occurs with flood plains.

6.1. Societal Vulnerabilities Highlighted by the 2011 Season

That we do not have to resort to types of extraordinary societal vulnerability to explain deaths in the 2011 tornado season does not imply that societal vulnerability played no role in the historic death toll. The determinants of fatalities in the recent past incorporate aspects of vulnerability, and our conclusion that extreme weather drove the death toll merely means that the number and strength of tornadoes in 2011 were sufficient, given societal vulnerabilities reflected in recent fatality patterns. We discussed societal vulnerabilities and the potential keys they offer for reducing tornado casualties in *Economic and Societal Impacts of Tornadoes* and will not repeat this entire discussion here. We do think, however, that the 2011 tornado season highlights several vulnerabilities worth noting as the nation tries to draw lessons from a remarkable year of severe-weather outbreaks.

Although more deaths occurred in permanent homes than in mobile homes in 2011, 21% of fatalities did occur in mobile homes, and since less than 10% of the U.S. population lives in mobile homes, the fatality rate even in 2011 was higher than in permanent homes. Thus efforts to reduce mobile-home casualties require continued attention. Shelters appear to offer cost-effective protection in tornado-prone states, and the HUD wind rule, though designed for hurricanes, may help reduce tornado fatalities (Simmons and Sutter 2009). Research shows that residents need sheltering options that they will be willing to undertake (Schmidlin et. al 2009). In addition, 16% of 2011 tornadoes occurred in what the Storm Prediction Center (SPC) describes as permanent buildings, which include businesses, schools, and churches. Only about 5% of fatalities since 1985 have occurred in businesses, and so this represents a way in which the 2011 season differed from recent experience. The elevated percentage of fatalities in these other types of buildings may be due to the number of violent tornadoes during the year.

Southeastern states are known to be vulnerable to tornadoes. The April 27 tornado outbreak did not fit the bill of the types of fatalities that occur disproportionately in the region—mobile-home fatalities and deaths in nighttime, in the fall and winter, and in weak and strong tornadoes. Yet, examination reveals that April tornadoes are almost as deadly in these states as winter tornadoes, and afternoon and evening tornadoes are more dangerous than in other parts of the country. The month of year and state or regional casualties effects may well have played a role in April fatalities. For instance, regression analysis suggests that expected fatalities are 31% higher for tornadoes in the Southeastern states, which suggests the April death toll might have been increased by 70 due to this vulnerability. The April 27 outbreak would have been extremely deadly wherever it occurred. The state and regional and especially time-of-year vulnerabilities are very large (larger

TABLE 6.1. Mobile- and Permanent-Home Fatalities per Tornado

EF-Scale	Mobile-Home Fatalities	Permanent-Home Fatalities	Ratio of Mobile- to Permanent-Home Fatalities
0	0.0001	0.0001	1.00
1	0.0090	0.0010	8.75
2	0.0655	0.0257	2.55
3	0.5249	0.2434	2.16
4	0.7424	0.9242	0.83
5	2.1667	12.833	0.17

For tornadoes in the 48 contiguous states, 1996–2007. Mobile- and permanent-home fatalities are expressed as rates per number of tornadoes in each EF-scale category.

in amplitude than the time-of-day vulnerability) and not very intuitive; it is not obvious why tornado lethality should vary substantially across the year. Consequently, these vulnerabilities need to be studied in detail in the future. For instance, do the regional or seasonal patterns observed for tornadoes extend to other types of extreme weather?

Violent tornadoes drove the deadly 2011 season as 81% of fatalities occurred in EF-4 and EF-5 tornadoes. The violent tornadoes are also responsible for the high proportion of fatalities (37% of fatalities, 45% of fatalities whose location is known as of this writing) relative to recent years in permanent homes. The incidence of violent tornadoes may also be behind the high proportion of business and other permanent-building fatalities during the year. Permanent-home fatalities remain primarily a product of violent tornadoes, as Table 6.1 highlights. Here we express mobile- and permanent-home fatalities over the years 1996 to 2007 per tornado for EF-scale categories. Only one fatality in each type of home occurred in EF-0 tornadoes, a rate of about one per 10,000 such tornadoes. Both rates increase as the EF-scale category increases, but the rate increases much faster for mobile homes and is more than double the permanent-home rate for EF-2 and EF-3 tornadoes, as the ratio in the final column of the table reports. More fatalities occur in permanent homes in EF-4 tornadoes by a small margin, with almost one fatality per tornado for each type of home. Finally, we observe more than 12 permanent-home fatalities per EF-5 tornado, which is about 6 times greater than for mobile homes. Alternatively, 54% of permanent-home fatalities occurred in violent tornadoes over these years, compared with 18% of mobile-home fatalities.

Violent tornadoes are extremely rare, and disseminating warnings so that people can shelter in an interior closet or bathroom normally affords adequate protection for permanent-home residents. And permanent homes actually provide pretty decent protection in an EF-5 tornado, because fatalities in these devastating tornadoes are often surprisingly low. First responders to Greensburg, Kansas, in 2007 expected that the EF-5 tornado that destroyed 95% of all buildings in the town would have killed hundreds, but the fatalities totaled 11. Given the extent of the destruction in the Joplin EF-5 tornado, it was almost miraculous that more lives were not lost.

The Joplin tornado does not reflect the worst-case scenario for permanent-home fatalities in violent tornadoes by any stretch of the imagination. About one-third of the population of the city of Joplin was in the tornado path, or an estimated 15,000 persons. Wurman et al. (2007) superimpose a potential long, wide-track violent tornado over different metropolitan areas to estimate potential urban impacts. They estimate that several hundred thousand residents could be in the path of such a tornado, so the number of persons in the path of an urban EF-5 tornado could exceed the 15,000 total in Joplin by an order of magnitude. We consequently expect that fatalities could exceed the toll in Joplin by an order magnitude if the permanent home–violent tornado issue is not addressed. The scenario proposed by Wurman et al. is a very low-probability event, but the reality exists that in the most violent tornadoes, sheltering in an interior room does not offer a high level of protection.

6.2. Can the Danger from Violent Tornadoes Be Efficiently Reduced?

Thus 2011 focuses our attention on the hard problem of permanent-home fatalities in violent tornadoes. One popular response to this problem, evident again this year, is to build tornado shelters and safe rooms. In the aftermath of the Alabama and Joplin tornadoes, FEMA hazard mitigation grant dollars will pay for scores of storm shelters. As we have previously discussed, shelters simply do not offer cost-effective protection for residents of permanent homes, because violent tornadoes are so rare and the homes offer good protection. We should not expect widespread adoption of such shelters, and we do not think it proper for experts who should know better to exaggerate tornado risk and frighten people into purchasing shelters that they would choose not to if they could dispassionately consider an accurate assessment of the risk.

Many discussions during the year also proposed extending lead times for tornado warnings. A variety of options offer the potential to increase the alert people receive for tornadoes, including dual-polarized Doppler radar, Phased Array Radar, a warn-on-forecast as opposed to a warn-on-tornado approach to warnings, and convective outlooks issued by the SPC. The potential exists through technology to extend warnings out to 30 minutes and to provide a warning based on forecast one or two hours in advance; convective outlooks are issued up to three days in advance, which indicates they are exhibiting increasing skill. The SPC convective outlook for the morning of April 27 had a high risk of severe weather and tornadoes across northern Mississippi and Alabama for later in the day, and the four EF-5 tornadoes that day occurred in this area.

Our research, however, suggests that we may have already reached the point of diminishing returns for tornado-warning lead time, given the current nature of warnings and warning response. Warnings do save lives, and longer lead times up to about 15 minutes reduce fatalities and injuries, with 15 minutes being close to the current average lead time. While lead times beyond 15 minutes reduce casualties relative to no warning, we have not found evidence of additional reductions in casualties due to increased lead times beyond 15 minutes. And we have also been unable to document life-saving effects for tornado watches—regression analysis suggests that a watch does not produce further reduction in casualties controlling for a warning. The results for watches are unlikely to change after 2011, since all but three of the year's 552 fatalities occurred in tornadoes that touched down within a valid tornado watch.

The lack of a marginal benefit of lead time beyond 15 minutes should perhaps come as no surprise. The time needed to respond to a warning is not long—it can take just a few seconds for most people to get to an interior room. Persons in special circumstances might benefit from additional lead time; for instance, elderly or handicapped residents will need longer to get to safety, while residents with small children or pets may need extra time to gather all the loved ones in a safe place. The value of lead time really arises through the opportunity to disseminate the warning, because people do not automatically receive a warning when it is issued. Theory does not tell us exactly how long it will take for warnings to reach residents, but the lack of reduction in casualties beyond a 15-minute lead time suggests that this is sufficient time for warnings to reach everyone who was going to act on it.

We are not opposed to efforts to extend warning lead times beyond 15 minutes or to improve convective outlooks, but we caution that researchers must be very careful to not oversell the benefits. As economists we interpret the world using economic models. Tornado warnings (like other types of weather forecasts) provide information to people, and in an economic model, information derives value from the actions it allows people to take (or not take). Meteorologists who wish to estimate the potential societal value of extending tornado-warning lead times should keep in mind this question: What action will people take with this information? If an extra 15 minutes or hour does not allow people to take an action that affords better protection than sheltering in a bathroom or closet or basement, the warning or convective outlook will fail to generate significant value to society.

Efforts to extend warning lead times or to improve convective outlooks need to keep in mind the localized nature of tornado threats and the value of equally localized tornado information. The average area of a tornado-damage path is less than half a square mile. A broad warning area quickly degrades the information value of a warning, as we have previously discussed (Simmons and Sutter 2011; Sutter and Erickson 2010). This helps explain why residents will often ignore tornado sirens but value street-level tracking offered by television meteorologists. The

National Weather Service (NWS) took a valuable step in the proper direction with the introduction of Storm Based Warnings (SBWs), which have the potential to reduce the area warned relative to county-based warnings by as much as 75%. Thirty- or 60-minute lead times on warnings achieved at the expense of increasing the area warned back to the county level are particularly unlikely to produce net benefits to society. Broad risk areas in convective outlooks will convey less information to residents.[1]

Extended lead times may hold the key to addressing the permanent home–violent tornado problem. Two basic alternatives exist for protecting people from natural hazards: evacuation and sheltering in place. We tend to use only one of the approaches for a given hazard. Earthquakes occur with virtually no warning and affect a large area, so residents need to shelter in place, and we try to harden targets—build structures and infrastructure that can survive an earthquake. For hurricanes the focus is on evacuation, as storm surge and the highest winds affect only the coastal area where the storm comes ashore. For tornadoes we generally think in terms of sheltering in place, which naturally (but we think erroneously) leads experts to see building tornado shelters everywhere as the way to reduce casualties. But violent tornadoes are just too rare to make shelters or safe rooms cost effective for residents of permanent homes. We need to think outside the box and switch our mental model for tornadoes from shelter in place to evacuation, at least in some circumstances. Permanent homes provide adequate protection against weak and strong tornadoes, but since building a safe room is not cost effective, residents need to get out of the path of violent tornadoes.

People already evacuate from tornadoes. Hammer and Schmidlin (2002) first documented this behavior during the May 3, 1999, Bridgecreek–Moore F5 tornado. None of the persons in their study

---

1. This discussion is focusing on benefits of warnings, watches, or convective outlooks for residents, which ultimately will be in the form of reduced casualties. Information might allow emergency managers, first responders, utilities, and others to prepare for a tornado outbreak and produce value by reducing noncasualty impacts.

who fled his or her home in cars in advance of that powerful tornado was killed or injured. Residents are undoubtedly fleeing for other tornadoes as well. Many reasonable objections could be raised against "evacuations" for tornadoes, and the objections illustrate the research meteorologists must undertake to allow evacuations to increase safety. Improved warnings could be integral in this effort. For instance, because permanent homes provide substantial protection, residents would face substantially greater risk if caught outside in a tornado than in their homes. And leaving the home increases the risk from accompanying thunderstorm threats. Thus permanent-home residents would need to know the strength of a tornado in real time to decide if they want to get out of the path of the tornado. It would also be valuable to forecast a violent tornado's strength and precise path. Emerging technologies like mobile Doppler radars, short-wavelength radars (McLaughlin et al. 2009), and phased-array radars may allow the observation of the lower levels of thunderstorms to measure the strength of a tornado (National Academy of Sciences 2002). Path forecasts would need to be more precise than current storm-based warnings so that evacuating residents do not move into the path and prevent too many people from attempting to flee and generate traffic jams. Longer warning lead times may allow residents to prepare to evacuate (e.g., arrange to go to a friend's or relative's house a safe distance from the tornado) and may be feasible in the future with warn-on-forecast for tornadoes (Stensrud et al. 2009).

Improved warnings might also address the mobile-home problem. Mobile homes face destruction at a lower wind speed threshold than permanent homes, reflected in the higher fatality rate at lower EF-scale levels depicted in Table 6.1. Thus, sheltering in place in a mobile home provides inadequate protection for a larger proportion of tornadoes, and so mobile-home residents will need to evacuate more often. Residents are reluctant to leave their homes, despite the fact that there are usually permanent buildings nearby that could offer protection (Schmdlin et al. 2009). Residents also seem disinterested in sheltering in a ditch during a tornado warning, which is reasonable given the likelihood any one home in an area under a warning will be struck and the discomfort (getting rained on during a thunderstorm) when their home is not struck by a tornado. And venturing outside as a tornado approaches runs counter to our natural impulse to put walls between us and a threat. A longer lead time with a more precise warning may offer a sufficient tradeoff of risk and time to prompt mobile-home residents to move to safer locations, perhaps the same type of evacuation that permanent-home residents could employ, or perhaps taking refuge in a structure offering greater protection.

The 2011 tornado season will likely be remembered as the year of long-track violent tornadoes. These tornadoes highlight the difficult residual problem of fatalities in permanent homes in the most powerful tornadoes. Addressing this vulnerability will require some out-of-the-box thinking and advances from meteorology to provide society with the information needed to make limited-scale evacuations for these most powerful tornadoes a reality.

REFERENCES

Chapter 1

Brooks, H. E., and C. A. Doswell III, 2002: Deaths in the 3 May 1999 Oklahoma City Tornado from a Historical Perspective. *Wea. Forecasting*, 17, 354–361.

Chapter 2

Ashley, W. S., 2007: Spatial and Temporal Analysis of Tornado Fatalities in the United States, 1880–2005. *Wea. Forecasting*, 22, 1214–1228.

Ashley, W. S., A. J. Krmenec, and R. Schwantes, 2008: Nocturnal Tornadoes. *Wea. Forecasting*, 23, 795–807.

Simmons, K. M., and D. Sutter, 2009: False Alarms, Tornado Warnings, and Tornado Casualties. *Weather, Climate and Society*, 1, 38–53.

Simmons, K. M., and D. Sutter. 2011. *Economic and Societal Impacts of Tornadoes*. Boston, MA: American Meteorological Society.

Chapter 3

Ashley, W. S., 2007: Spatial and Temporal Analysis of Tornado Fatalities in the United States, 1880–2005. *Wea. Forecasting*, **22**, 1214–1228.

Boruff, B. J., J. A. Easoz, S. D. Jones, H. R. Landry, J. D. Mitchem, and S. L. Cutter, 2003: Tornado Hazards in the United States. *Climate Research*, **24**, 103-117.

Downton, M. W., J. Z. B. Miller, and R. A. Pielke, Jr., 2005: Reanalysis of the U.S. national flood loss database. *Nat. Hazards Rev.*, **6**, 13–22.

Gall, M., K. A. Borden, and S. L. Cutter, 2009: When Do Losses Count? Six Fallacies of Natural Hazard Loss Data. *Bull. Amer. Meteor. Soc.*, **90**, 799–809.

Mileti, D. S. 1999. *Disasters by Design: A Reassessment of Natural Hazards in the United States.* Washington, DC: Joseph Henry Press.

Simmons, K. M., and D. Sutter, 2008: Manufactured Home Building Regulations and the February 2, 2007 Florida Tornadoes. *Nat. Hazards*, **46**, 415–425.

Simmons, K. M., and D. Sutter, 2009: False Alarms, Tornado Warnings, and Tornado Casualties. *Weather, Climate and Society*, **1**, 38–53.

Simmons, K. M., and D. Sutter. 2011. *Economic and Societal Impacts of Tornadoes.* Boston, MA: American Meteorological Society.

Chapter 4

AL.com http://blog.al.com/spotnews/2011/04/alabama_tornadoes_alabama_power.html

AL.com http://blog.al.com/spotnews/2011/09/alabama_tornadoes_early_power.html

Bieringer, P., and P. S. Ray, 1994: A Comparison of Tornado Warning Lead Times with and without NEXRAD Doppler Radar. *Wea. Forecasting*, **11**, 47–52.

Friday, E. W., Jr., 1994: The Modernization and Associated Restructuring of the National Weather Service: An Overview. *Bull. Amer. Meteor. Soc.*, **75**, 43–52.

Polger, P. D., B. S. Goldsmith, R. C. Pryzwarty, and J. R. Bocchierri, 1994: National Weather Service Warning Performance Based on the WSR-88D. *Bull. Amer. Meteor. Soc.*, **75**, 203–214.

Samenow, J., 2011. "Storm Warning Expert: Power Outages a Principal Cause of High Alabama Death Toll." http://www.washingtonpost.com/blogs/capital-weather-gang/post/storm-warning-expert-power-outages-a-principle-cause-of-high-alabama-tornado-death-toll/2011/05/05/AFWS8QzF_blog.html.

Simmons, K. M., and D. Sutter. 2011. *Economic and Societal Impacts of Tornadoes*. Boston, MA: American Meteorological Society.

Chapter 5

Belasen, A. R., and S. W. Polachek, 2009: How Disasters Affect Local Labor Markets: The Effects of Hurricanes in Florida. *Journal of Human Resources*, University of Wisconsin Press, vol. 44(1).

Burby, R. J., Ed. 1998. *Cooperating with Nature*. Washington, DC: Joseph Henry Press.

Chamlee-Wright, E., 2010. *The Cultural and Political Economy of Recovery: Social Learning in a Post-Disaster Environment*. New York: Routledge.

Chamlee-Wright, E., and V. H. Storr, 2009: Club Goods and Post-Disaster Community Return. *Rationality and Society*, 21(4), 429–458.

Ewing, B., J. B. Kruse, and D. Sutter, 2007: Hurricanes and Economic Research: An Introduction to the Hurricane Katrina Symposium. *Southern Economic Journal*, 74(2), 315–325.

Ewing, B., J. B. Kruse, and M. Thompson, 2003: A Comparison of Employment Growth and Stability Before and After the Fort Worth Tornado. *Environ. Hazards*, 5, 83–91.

Ewing, B., J. B. Kruse, and M. Thompson, 2005: An Empirical Examination of the Corpus Christi Unemployment Rate and Hurricane Bret. *Nat. Hazards Rev.*, 4, 191–196.

Ewing, B., J. B. Kruse, and M. A. Thompson, 2007: Twister! Employment Responses to the 3 May 1999 Oklahoma City Tornado. *Applied Economics*, iFirst, 1–12.

Ewing, B., J. B. Kruse, and Y. Wang, 2007: Local Housing Price Index Analysis in Wind-Disaster-Prone Areas. *Nat. Hazards*, 40, 463–483.

Grazulis, T. P., 1993. *Significant Tornadoes 1680–1991*. Environmental Films, St. Johnsbury, VT.

Larson, E., 1999. *Isaac's Storm*. New York: Vintage Books.

Marshment, R., and C. Rogers, 2000: Measuring Highway Bypass Impacts on Small Town Business Districts. *Review of Urban and Regional Development Studies*, 30(3), 315–330.

Mileti, D. S. 1999. *Disasters by Design: A Reassessment of Natural Hazards in the United States.* Washington, DC: Joseph Henry Press.

Mill, J. 1848. *Principles of Political Economy.* New York: A M Kelley Publishers.

Sobel, R. S., and P. T. Leeson, 2007: The Use of Knowledge in Natural Disaster Relief Management. *Independent Review*, 11(4), 519–532.

Chapter 6

Hammer, B., and T. W. Schmidlin, 2002: Response to Warnings during the 3 May 1999 Oklahoma City Tornado: Reasons and Relative Injury Rates. *Wea. Forecasting*, 17, 577–581.

McLaughlin, D., et al., 2009: Short-wavelength technology and the potential for distributed networks of small radar systems. *Bull. Amer. Meteor. Soc.*, 90, 1797–1817.

National Academy of Sciences. 2002. *Weather Radar Technology Beyond NEXRAD.* National Academies Press (available online at: www.nap.edu/openbook/0309084660/html/1.html).

Schmidlin, T. W., B. O. Hammer, Y. Ono, and P. S. King, 2009: Tornado Shelter-Seeking Behavior and Tornado Shelter Options Among Mobile Home Residents in the United States. *Nat. Hazards*, 48, 191–201.

Simmons, K. M., and D. Sutter, 2009: False Alarms, Tornado Warnings, and Tornado Casualties. *Weather, Climate and Society*, 1, 38–53.

Simmons, K. M., and D. Sutter. 2011. *Economic and Societal Impacts of Tornadoes.* Boston, MA: American Meteorological Society.

Stensrud, D. J., et al., 2009: Convective-scale warn on forecast: A vision for 2020. *Bull. Amer. Meteor. Soc.*, 90, 1487–1499.

Sutter, D., and S. Erickson, 2010: The Time Cost of Tornado Warnings and the Savings with Storm Based Warnings. *Weather, Climate and Society*, 2, 103–112.

Wurman, J., C. Alexander, P. Robinson, and Y. Richardson, 2007: Low-level winds in tornadoes and potential catastrophic tornado impacts in urban areas. *Bull. Amer. Meteor. Soc.*, 88, 31–46.

INDEX